AF388893

Spatial Heterogeneity of Forest-Steppes

Spatial Heterogeneity of Forest-Steppes

Editor

Laszlo Erdos

MDPI • Basel • Beijing • Wuhan • Barcelona • Belgrade • Manchester • Tokyo • Cluj • Tianjin

Editor
Laszlo Erdos
IEB Department of Terrestrial Ecology,
MTA Centre Ecology Research
Hungary

Editorial Office
MDPI
St. Alban-Anlage 66
4052 Basel, Switzerland

This is a reprint of articles from the Special Issue published online in the open access journal *Forests* (ISSN 1999-4907) (available at: https://www.mdpi.com/journal/forests/special_issues/Forest_Steppes).

For citation purposes, cite each article independently as indicated on the article page online and as indicated below:

LastName, A.A.; LastName, B.B.; LastName, C.C. Article Title. *Journal Name* **Year**, *Volume Number*, Page Range.

ISBN 978-3-0365-0670-8 (Hbk)
ISBN 978-3-0365-0671-5 (PDF)

Cover image courtesy of László Erdős.

Contents

About the Editor

Laszlo Erdos graduated from the University of Szeged (Hungary) as a biologist in 2005. He earned his PhD in vegetation ecology at the University of Pécs (Hungary). His main area of interest is the vegetation ecology of Eurasian forest-steppes. Currently, he is a research fellow at the Centre for Ecological Research (Vácrátót) and at the University of Debrecen (Debrecen). He also serves as visiting lecturer at the Eötvös Loránd University (Budapest).

Preface to "Spatial Heterogeneity of Forest-Steppes"

Situated between the closed-canopy forests and the open steppes, forest-steppes feature a highly complex mosaic of woody and herbaceous habitats. Though forest-steppes exits in North and South-America, the largest forest-steppes are found in Eurasia, where they extend as an almost continuous belt from the Carpathian Basin in Eastern Central Europe to the Far East in China and Russia. Despite its tremendous biogeographical, ecological and even economic importance, the forest-steppe zone has received surprisingly little scientific attention. As forest-steppe landscapes are composed of a high variety of different forest, scrub and grassland communities and the ecotones among them, spatial heterogeneity is a key issue if we are to understand how forest-steppe ecosystems work and how they may react to environmental changes at the local, regional, and global scales.

Forest-steppes exist in a transitional semi-humid to semi-arid climate that enables the long-term coexistence of forests and grasslands, as these have roughly equal chances to develop. The competition outcome at every given site depends on minor differences in terrain, soil and microclimate characteristics. Once established, grassland species probably modify the environment to favor the continued existence of the grassland patch, while trees and shrubs support the persistence of woody vegetation. This seems to stabilize forest-steppe mosaics, though modifications in climate, herbivore pressure or fire frequency and intensity may alter the balance. Unfortunately, it is not known how strong these feedbacks are, thus we do not know if and where the current patterns are able to resist natural and anthropogenic changes.

Forest-steppes host exceptional taxonomic diversity from local to landscape to regional scales. In addition, they harbor a large number of rare, nationally protected, and internationally threatened species. Moreover, the functional diversity of forest-steppes is very high, with a large variety of life-forms.

Forest-steppes provide us with a number of vital ecosystem services, including a large carbon sequestration capacity, protection from soil erosion, and aesthetic value.

Unfortunately, large forest-steppe areas, especially in the most populous regions, have been destroyed and replaced with (mostly non-native) tree plantations or arable fields, while the remaining near-natural areas are threatened by the spread of invasive species, fragmentation, and overexploitation. Thus, preserving forest-steppe ecosystems presents a great challenge to conservationists. May we hope that a better understanding of this unique ecosystem will contribute to its more efficient conservation.

Laszlo Erdos
Editor

Article

Fine-Scale Microclimate Pattern in Forest-Steppe Habitat

Gabriella Süle [1,*], János Balogh [1], Szilvia Fóti [1,2], Bernadett Gecse [1] and László Körmöczi [3]

[1]   Department of Plant Physiology and Plant Ecology, Institute of Biological Sciences, Szent István University, H-2100 Gödöllő, Hungary; balogh.janos@szie.hu (J.B.); Foti.Szilvia@szie.hu (S.F.); gecsebernadett@gmail.com (B.G.)

[2]   MTA-SZIE Agroecology Research Group, Szent István University, H-2100 Gödöllő, Hungary

[3]   Department of Ecology, Faculty of Science and Informatics, University of Szeged, 6720 Szeged, Hungary; kormoczi@bio.u-szeged.hu

[*]   Correspondence: sulegaby@gmail.com; Tel.: +36-70-593-2870

Received: 12 September 2020; Accepted: 5 October 2020; Published: 9 October 2020

Abstract: Microclimate and vegetation architecture are interdependent. Little information is available, however, about the fine-scale spatio-temporal relationship between the microclimate and herb layer of forest-steppe mosaics. In 2018 a three-season-long vegetation sampling and measurements of air temperature and air humidity were performed along 4 transects (44 m long each) in the herb layer with 89 dataloggers in the sandy region of Central Hungary, in a poplar grove and the surrounding open grassland. In order to improve data analysis, we introduced the use of a duration curve widely used in hydrology and proved to be useful in the processing of intensive climatic data. We analysed the effect of the direction and altitude of the solar irradiation and the edge effect on the microclimatic pattern. We also surveyed, seasonally, the spatial pattern of the exceedance rate for the vapour pressure deficit (VPD) in relation to the transect direction and to the edge of the grove. The exceedance rate for the VPD indicated considerable seasonal differences. The VPD exceedance rate indicates the stress effect for the vegetation. The moderating effect of the grove was small at 1.2 kPa VPD, but at 3.0 kPa—stronger stress—it was considerable. On the warmer side of the transects, mostly exposed at the south-eastern edge, the exceedance rate rose abruptly with distance from the edge compared to the gradual increase on the colder side. The cardinal and intercardinal directions as well as the altitude of the Sun all had influences on the moderating and shading effects of the grove. The southern edge was not always consistently the warmest. The distribution of the VPD values above the 3.0 kPa threshold varied within a seemingly homogeneous grassland, which highlights the importance of fine-scale sampling and analysis. This knowledge is valuable for assessing the dynamics and spatio-temporal patterns of abiotic factors and physiognomy in this type of ecosystem.

Keywords: duration curve; forest steppe; sensor network; spatio-temporal microclimate pattern; temperature–humidity data logger; vapour pressure deficit

---

1. Introduction

Climate change is expected to have a major impact on long-term community dynamics worldwide. Among other things, the global temperature increase and extreme weather events have a significant impact on the structure of forest ecosystems and also on the dynamics of environmental factors in forest patches and nearby open areas [1–3]. Habitats with different attributes (e.g., vegetation physiognomy and plant species composition) may respond differently to the changes, depending on their sensitivity and resilience. Transition zones could be sensitive habitats, where the structures of two different vegetation types have significant effects on each other, while their strong biotic and abiotic relationships determine the dynamics of the vegetation [4]. Typical transition zones are forest fragments, which, in many cases, are human-induced [5].

Studies about fragmented habitats have mainly focused on the effects of the surrounding open areas around the forest fragments [6–8]. However, there are naturally fragmented vegetation types, such as forest-steppe habitats in Central Hungary. In the case of this sandy forest-steppe vegetation, the fragmented structure has a natural origin. In this habitat, natural drying processes have been observed for decades [9,10], inhibiting the growth of woody vegetation. In addition to these processes, the seeds of tree species in this habitat can only germinate in depressions with ideal topographic conditions where a sufficient amount of water can accumulate. However, these seedlings can rarely develop into even loosely closed tree groups. Thus, the succession dynamics in a sandy-forest-steppe habitat are rather slow due to the vegetation edges, where the strong abiotic differences between the grassland and the fragment prevent forest expansion and development [10]. The sandy forest-steppe is a vegetation type that is in danger of total extinction in the near future due to the aridification in the Pannonian region [11]. Thus, improving our knowledge of this type of ecosystem is important for understanding the dynamics of abiotic and biotic factors in transition zones. Therefore, it is important to examine the effect of smaller groups of trees and larger forest patches on the surrounding grassland matrix where an edge effect is observed [12–14].

Among the environmental factors that characterize a habitat, the microclimate strongly influences the growth and distribution of plant species [15–18]. The most important microclimatic components that are usually examined included the air temperature, air humidity, wind force, and solar radiation [4,16]. In a study by Lin and Lin [19], environmental measurements showed that solar radiation and wind velocity had significant opposite effects on air temperature in urban vegetation. The intensity of solar radiation had a positive effect, while wind velocity had a negative effect. Thus, the cooling effect of woody vegetation is highly dependent on these factors. Hence, temperature and humidity are the main variables, from which we can also obtain the vapour pressure deficit [4,20,21]. The vapour pressure of the air in the forest and in the open areas would be similar if the vapour pressure deficit and relative humidity were determined by the air temperature. The difference in the air temperature causes the differences between the relative humidity of open and canopy-covered areas [17,22,23]. The vapour pressure deficit (VPD) can be an important limiting factor in plant growth because conditions with above-threshold values can be considered stressful. Under conditions with above-threshold values, photosynthetic CO_2 uptake is strongly limited due to stomatal closure to prevent water loss [5,22,24,25].

These variables are also closely related to each other and are influenced by aspects of the vegetation structure such as the height, canopy cover, species composition, etc. [1,26]. The canopy cover directly influences the below-canopy and nearby microclimate, decreasing the solar radiation energy reaching the surface during the day while retaining outgoing longwave radiation during the night [1,26,27]. In an Atlantic forest fragment, the transient nature of the edge was observed by microclimate measurements, where the air temperature at the edge was higher than in the forest interior, while the relative humidity was lower [27]. This transient nature was also reported in a Douglas-fir forest in the U.S. [13]. Therefore, the forested areas heat up less during daytime and cool down less during night-time. In the UK, in a study of a semi-natural temperate forest, the air temperature in the forest was lower than that of the grassland and decreased from the canopy to the understory on a sunny summer day. The largest difference in air temperature between these two different areas was approximately 3 °C [12]. These microclimatic differences between open and canopy-covered areas may be much larger and more relevant, depending on the geographical location and physiognomy of the vegetation studied. Thus, little information is available about the spatio-temporal relationship between the microclimate components, the vegetation structure and ecosystem functioning in fragmented vegetations, although this knowledge is important for understanding the ecological functioning of forest ecosystems.

The edge effect is the microclimatic or vegetation structural difference between the forest edge and the interior of the forest [5]. The size of the forest patch considerably influences the extent of the edge effect and the difference between the forest interior and peripheral areas such as nearby open grassland [4,14]. According to Murcia's review [14], there are three types of edge effects on the forest fragments. The first one is the abiotic effect, in which a change in the environmental conditions will create a habitat matrix with physical conditions different from the forest (e.g., differences in structural complexity and biomass). The two factors influencing the abiotic effects include the physiognomy and orientation. The second is the direct biological effect, in which there is a change in the abundance and distribution of species in the edge area because of differences in species tolerances. This is due to the sudden changes in the physical conditions in the edges, which may increase the tree mortality caused by wind force (wind throw) or by fire. This will lead to a change in species composition. The third is the indirect biological effect: alterations in species interactions, such as competition, predation, pollination, etc.

The stages of natural succession modify the edges during the development time; therefore, fluctuations in the distribution of plant species can be observed [28–30]. The current vegetation pattern corresponds to the present environmental conditions. The duration of the study is a very important parameter for such research topics, and also, the scale of sampling is relevant, but fine-scale studies are very rare [31]. Most studies about forest fragments and transition zones have used large-scale resolutions with a minimum of 20–50 m intervals between the measurement points, and the vegetation sampling has usually consisted of random, non-contiguous plots. As a result, information on fine-scale vegetation patterns is lost, and significant microclimatic differences may not be detected in the transition zone [5,13,20]. The fine-scale resolution with a sensor network helps to explore small-but-significant differences within the vegetation [31].

The main goal of this article was to assess the relationship between the microclimate and the microcoenological structure of the herb layer and to describe the microclimate-modifying effect of a grove in a sandy forest-steppe habitat. In three phenological stages of the one-year-long vegetation period, we analysed the air vapour pressure deficit (VPD) and plant species composition of the herb layer in four intersecting transects with different cardinal directions through a group of trees. The air temperature and humidity were measured in the herb layer at a fine spatial and temporal scale with 89 devices during the measurement campaigns, and 350 microcoenological relevés were made per season. We hypothesized that the VPD-modifying effect of the grove would gradually decrease in all directions away from the edge. We also assumed that the spatial microclimate patterns did not differ from season to season, that only the intensity of the modifying effect changed. The response of vegetation to the microclimate pattern was quantified according to the distribution of ecological indicator values. Our third hypothesis was that the coenological and indication structure of the herb layer differed only between the grove and the open areas and they were homogeneous in the open grassland.

2. Materials and Methods

2.1. Study Area

Our study site (Figure 1) was situated in Central Hungary (Fülöpháza region of the Kiskunság National Park; 46°53′28.18″ N 19°24′46.91″ E., 107 m a.s.l.). Throughout history, this region has been heavily cultivated, and quicksand was stopped by afforestation with invasive tree species. These processes destroyed much of the sandy forest-steppe habitat and gave space to both woody and herbaceous invasive species [10,32]. Natural forests have remained in the landscape in relatively small patches, and the sandy grasslands have, in many cases, been regenerated by secondary succession. Thus, the maintenance and restoration of the remaining natural areas is extremely important in this region. This habitat is characterized by a semi-desert climate, well indicated by the xerophytic dominant grass species, which, in the primary grasslands, are *Festuca vaginata* Waldst. & Kit. ex Willd. or *Festuca rupicola* Heuff., while the secondary grasslands contain mostly *Bromus tectorum* L. and *Secale sylvestre* Host. The dominant tree species in the natural forests are *Populus alba* L. or *Quercus robur* L., while the forest plantations are mostly dominated by *Robinia pseudoacacia* L. or *Pinus nigra*

J.F. Arnold [32]. A grove of poplar (*Populus alba*) and the surrounding grassland were selected for this study. Measurements of microclimate components and vegetation sampling were performed in four intersecting transects (44 m long each) with different cardinal directions, forming, together, a star-shape sampling arrangement with the group of trees in the middle. The diameter of the tree group was 15 m on average.

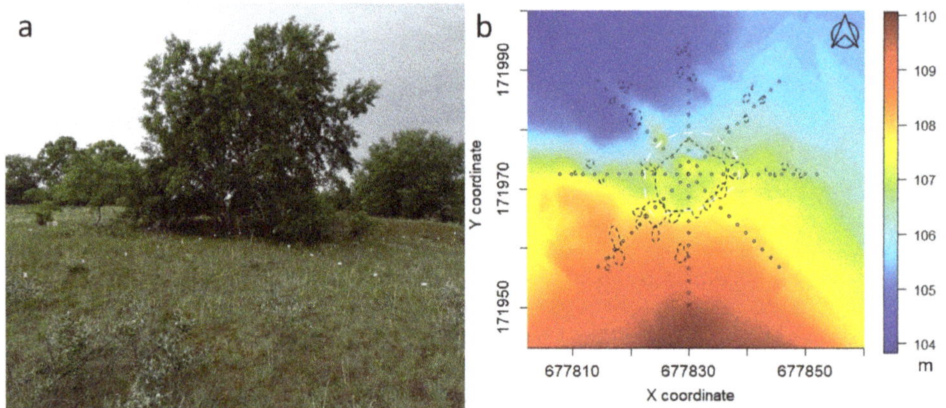

Figure 1. Study site (**a**) and relief map with the indication of the grove (**b**). Coordinates refer to the Hungarian Unified National Projection System. There were 89 measurement positions (4 × 22 positions in the four transects and the centre as position 12 in all transects). White dashed line: visual tree edge. Black dashed lines: the position and extent of individual shrubs and poplar sprouts.

2.2. Microclimate Measurements

Air temperature and air humidity were measured in the herb layer with a sensor network for 48 h (1-min resolution) during three measurement campaigns in different phenological stages of the vegetation in 2018 (May, July, and October). The data loggers were placed 20 cm above the soil surface, at the average height of the herbaceous vegetation, along the transects in 2 m intervals (23 measuring positions in each transect). The Crossbow MICA XM2110CA mote (Crossbow Technology Inc., Milpitas, CA, USA), UNI-T UT330B Mini USB Temperature Humidity logger (UNI-TREND Technology Co. Ltd., Guangdong, China), and Voltcraft DL-120TH USB Temperature Humidity logger (Voltcraft, Hirschau, Bavaria, Germany) were used in a sensor network, including 89 dataloggers altogether. The sensors were shielded with a white plastic plate to avoid solar radiation heating. Before the measurements, the sensors were calibrated. We selected precipitation-free measurement periods, but the sky was cloudy during the observation in May. The main changes in the weather were recorded (e.g., clouds' shading and movement). The location of the visual edges of the grove, the positions of bushes and trees in the surrounding area, and the shadow of the grove were also recorded.

2.3. Vegetation Sampling

Vegetation data were also collected along the transects. Microcoenological relevés were recorded in 0.5 m × 0.5 m contiguous plots in each season in parallel with the micrometeorological measurements. Plant names and indicator values (TZ temperature requirement; WZ moisture requirement elaborated by Zólyomi in Table S1) were used according to FLORA Database 1.2. These ecological indicator values quantify the environmental optimums for plant species based on their occurrences in natural habitats. Lower indicator values mean lower temperature and moisture requirements, while higher values mean higher requirements [33].

2.4. Vapour Pressure Deficit and Duration Curve Method

The vapour pressure deficit was computed from the relative air humidity (RH) and air temperature (t) according to the formula developed by Bolton [21]:

$$VPD = (100 - RH) \times 6.112 \times e\char94(17.67 \times t/(t + 234.5))$$

(1)

with *t* in °C, *RH* in %, and *VPD* in Pa.

In hydrology, the "flow duration curve" is a widely used method for detecting the rate of occurrence of values for a variable above a certain critical limit (flooding degree in hydrology). With the help of this method, one can identify the duration of a flood, which means the number of days flooded during the current period [34,35]. Since micrometeorological data are large sets of fluctuating time series, similar to the hydrological data, we consider the duration curve to be a promising tool for the analysis of temperature, humidity, or vapour pressure deficit data in plant ecological studies. Our data have a diurnal cycle and may be of any temporal resolution. Our study focused primarily on the derived data, such as the percentage of the VPD values above an appropriate threshold (1.2 or 3.0 kPa) over a 24-h period (exceedance rate) that can indicate the microclimatic conditions of the vegetation. A VPD duration curve (DC) was constructed from a 24-h period of records, from 12:00 to 12:00 in each measuring position. The DC of one variable was created by sorting all the data in descending order. Thus, the rank of the highest value was 1, while that of the smallest was n (number of measurements). The ordered data can be plotted to show the DC, where the relative order (e.g., percentage) on the X-axis reflects the exceedance probability for a particular value of the variable on the Y-axis (e.g., the vapour pressure deficit at one measuring position), indicating the percentage of time a given value was equalled or exceeded over the measurement period. The tendency of the curve shows the relationship between the exceedance probabilities and the examined variable. This graph is called a period-of-record DC. Based on this method, the exceeded values can be easily determined over the measurement period [34,36].

2.5. Data Processing

Data processing was carried out on the temperature and humidity data recorded at a per-minute frequency; during 2018, altogether, more than 1,200,000 records were processed. A 24-h recording period was used to calculate the VPD and VPD exceedance rates (%) for each measurement position between 12:00 and 12:00. Statistical evaluation was performed in R [37]: VPD calculation, the spline interpolation of spatial plots (akima package) [38], principal coordinate analysis (PCoA; vegan package) [39], boxplots (calculation of average, median, min, and max), and data visualization (ggplot2 package) [40].

The principal coordinate analysis (PCoA) ordinations were made based on microcoenological data and the DCs of VPD values. In the latter case, the ordination was practically performed on the quantiles of the VPD (1% resolution). Spatial maps were generated by spline interpolation to plot the VPD data for the whole study site. The interpolation error can be smaller in this case than that when using other interpolations [41].

3. Results

The distribution of the VPD values originating from 89 locations (4 × 22 positions in the four transects and the centre position) representing three seasons is illustrated by boxplots (Figure 2). The visually identified edges of the grove are also marked in the figures to identify the below-canopy area, the edges, and the open areas. In the figures, the left sides represent the colder sides and the right sides, the warmer ones, as determined by the irradiation and shadow patterns (cardinal directions).

During a 24-h period, differences between the grassland, edge, and below-canopy areas can be determined based on the vapour pressure deficit. The below-canopy VPD values were consistently lower than the edge or grassland VPD values. Except for the summer measurements, the below-canopy local average VPD values were lower than the threshold. The differences between the local averages and the threshold were considerable in the grassland, mostly in July. The medians of the values were very low in May and October, whereas in July, the median was mostly above the threshold. In July, the median and mean values of the below-canopy positions were very similar to each other and to the threshold. These three derivatives did not differ considerably from one another during the summer measurements (Figure 2).

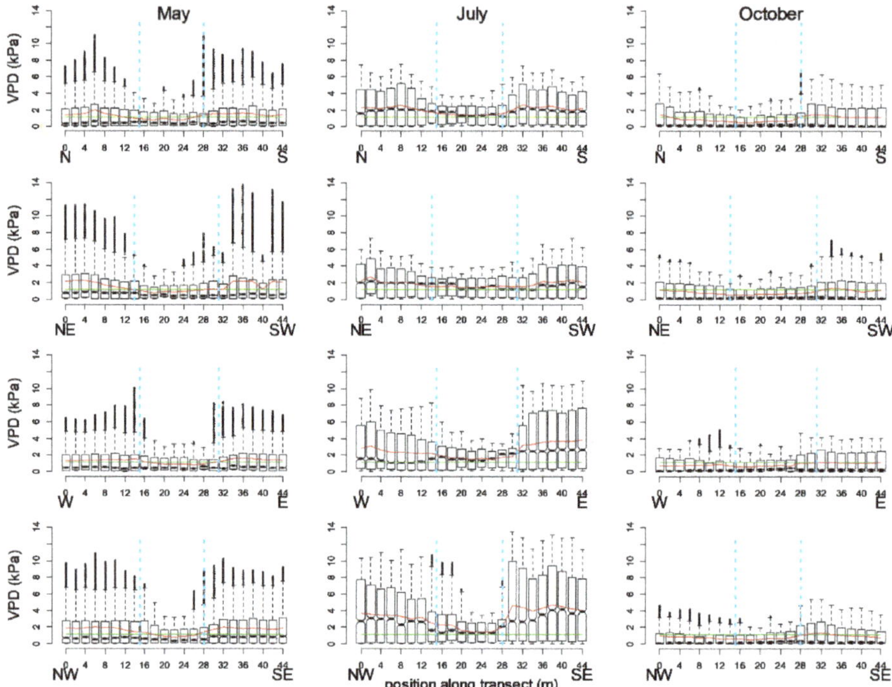

Figure 2. Boxplots of vapour pressure deficit (VPD) values from 24-h periods of three measurement campaigns according to the cardinal directions. Blue dashed vertical lines: visual tree edges; green horizontal lines: VPD threshold (1.2 kPa); red lines: local averages.

The pattern of micrometeorological parameters was determined by the exposure and the position (distance) to the grove in each season. As with temperature fluctuations, the range of VPD was significantly smaller within the grove than in the open area. In the edges, we observed a special behaviour of the VPD: in the warmer edge, which was more exposed to irradiation, the values were higher with a more abrupt rise than in the open area. On the colder side of the transects, this elevated VPD did not occur at the edges, and the deficit increased more smoothly with distance from the tree group.

The durations of the values being over the thresholds provide important information about the spatio-temporal pattern of the VPD. Figure 3 compares the DCs of the two ends, the edges, and the centre of the SE–NW (Southeast–Northwest) transect for the three measurement periods. Among the four transects, the difference in the effect of the exposure was the most pronounced in this one, but the behaviour of the other three transects was similar. In terms of the stress threshold (1.2 kPa) exceedance rates, the summer patterns differed significantly from the spring and autumn. The exceedance rates ranged from 29% to 41% in May, from 52% to 60% in July, and from 22% to 36% in October. In terms of VPD distribution, however, further groupings emerged. The distributions of the values measured in the middle of the grove showed relatively small seasonal differences, with longer but not stronger exceedance rates measured during the summer. The exceedance rates were 29% in May, 52% in July, and 29% in October. Although the exceedance rate for 1.2 kPa showed a small variation in each measurement period, the intensity of the exceedance already differed significantly between the individual measurement positions. The duration curves of both the end-of-transect and edge measurement series ran close to each other in each period, but their distributions differed significantly. The maxima of the end-of-transect and edge series for the spring and summer measurements were in the range of 8–11 kPa, and the others were in the range of 3–5 kPa. The spring and summer measurement series ran close together, except for the summer end-of-transect (open grassland) measurement series (Figure 3). The VPDs of the latter two series of measurements, on the other hand, were well above the others, with a more-than-45% exceedance rate with respect to 3.0 kPa, while the rates for the others were below 25%. Above 2.5 kPa, the variability of the exceedance rates began to increase in all transects; therefore, we examined the exceedance at 3.0 kPa as well (Figures 3 and 4b). The exceedance rates with respect to 3.0 kPa ranged from 1% to 25% in May, from 6% to 53% in July, and from 1% to 16% in October.

The exceedance rate of the autumn values with respect to the stress threshold did not differ significantly from that of the spring values of cloudy weather, but despite the sunny and warm daytime weather, the distributions above the critical limit were completely different. The maximum values did not exceed 5 kPa, and they were only slightly higher than the values below canopy.

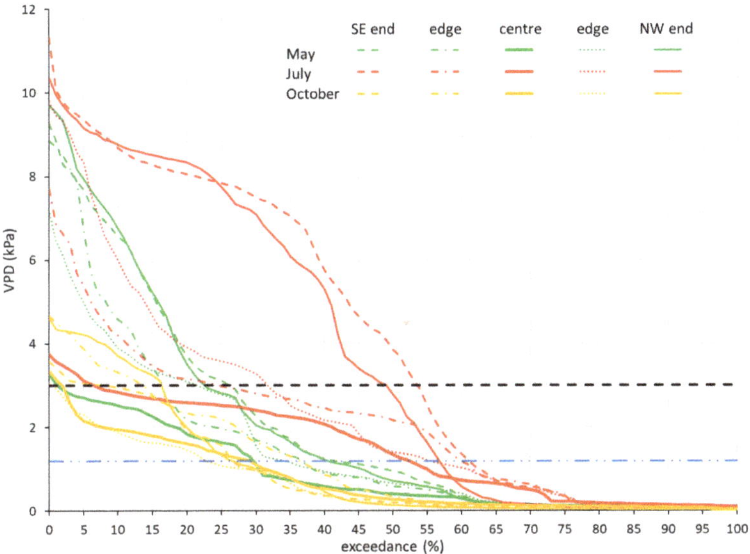

Figure 3. VPD duration curves for the SE–NW transect from three 24-h measurement periods. Blue dashed line indicates the 1.2 kPa physiological threshold; black dashed line indicates 3.0 kPa threshold, above which the exceedance rates significantly diversified.

The exceedance rate with respect to the threshold of 1.2 kPa calculated from the VPD DCs also showed seasonal variability (Figure 4a). In each of the three measurement periods, at least 20% of the values were above the threshold. In the spring and autumn measurements, the exceedance rate was 20–45%, while in the summer period, it was 48–62%. Spring and autumn did not differ significantly from each other. The below-canopy exceedance rate was balanced and tended to be low, but the exceedance rate varied more across open areas. We also found differences in the VPD exceedance rate between the opposite ends of the transects. Choosing 3.0 kPa as the threshold value for the exceedance rate, the curves show a stronger difference between the areas in the opposite edges of the grove (Figure 4b). On sunny days, on the colder side, the lower values were even more pronounced when the Sun was at a lower altitude (October) at noon, with, sometimes, a 0–10% duration of values being above 3.0 kPa, as opposed to a 25–30% duration on the warmer side (Figure 5).

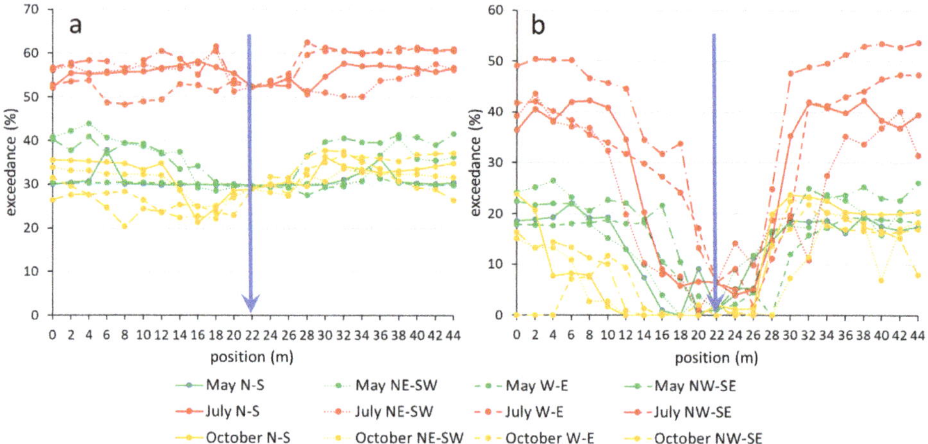

Figure 4. Exceedance (duration) (%) for VPD values above 1.2 kPa (**a**) and VPD values above 3.0 kPa (**b**) from 24-h period of three measurement periods; the left sides of the positions are the "cold" ends (W, NW, N, and NE), the right sides are the "warm" ends (E, SE, S, and SW), and the blue arrows are the centres of the transects.

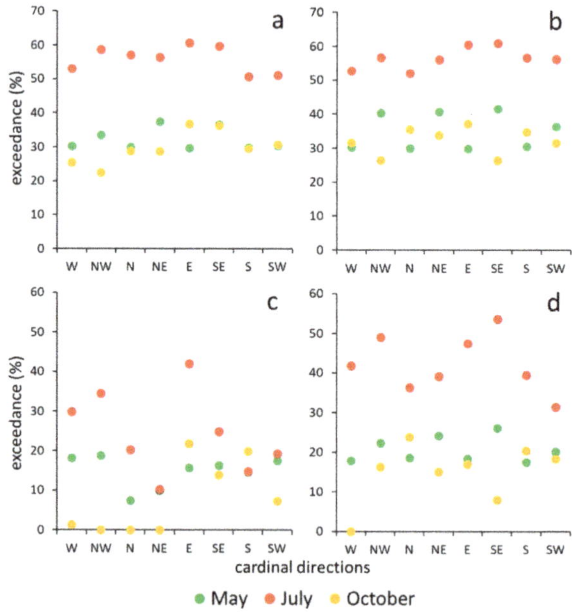

Figure 5. Exceedance rates for VPD values with respect to 1.2 kPa (top panels: (**a**,**b**)) and 3.0 kPa (lower panels: (**c**,**d**)) from 24-h period of three measurement periods; (**a**,**c**): edge positions; (**b**,**d**): grassland end positions.

There were no remarkable differences among (all of) the edges during the spring measurements. In the case of the exceedance rate for the VPD values with respect to 1.2 kPa (Figure 5a,b), in the edges, the highest values always occurred on the eastern and south-eastern side, as well as in the grassland end positions, except for in October. In the edges, an increasing tendency could be observed from west to south-east, followed by a remarkable decreasing trend from south to west, although these trends did not appear in the case of grassland end positions. The values in the grassland varied during the three measurement periods. However, in the case of the exceedance rate of the VPD values with respect to 3.0 kPa (Figure 5c,d), the tendency was quite different in the edges. In the NW (Northwest), N (North), and NE (Northeast) edges, the exceedance rate was 0% with respect to 3.0 kPa. This tendency was also observed in May and July, but the fall was not as strong as in the October data.

Depending on the weather conditions, the spatial pattern of the exceedance rate (Figure 6) also varied. At the 3.0 kPa threshold, the microclimatic differences among the three seasons were striking; the high VPD displayed the areas with consistently highly stressful conditions. In each of the three measurement periods, the below-canopy positions had lower values, but the values of the edges and the surrounding grassland were varied. There was only a slight difference between the open areas on the different sides of the tree group (Figure 6a). In October (Figure 6c), the transition between the warmer (E–SE (East–Southeast)) and colder (W–NW (west– Northwest)) sides was more gradual than in July (Figure 6b). In both of the later cases (summer and autumn), the measurements were made under clear skies. In July, the opposite grassland sides of the grove were in sync, NE and SW had lower values, and SE and NW had higher values. On the other hand, in October, the warmer and colder zones expanded considerably, and the transition between the two zones was sharp. In October, the warmer grassland side was E–SW, and W–NE was notably colder. In both periods, the eastern and south-eastern open areas had the highest values.

The ordinations also showed the different behaviour of the open areas of the SE–NW and E–W transects. During July, the open areas of these two transects were clearly distinct from each other, in contrast to the other two transects, where the confidence ellipses largely overlapped. The below-canopy positions were separated only in the SE–NW and E–W transects, while no outstanding differences were observed in the microclimatic conditions of the two other transects (Figure 7).

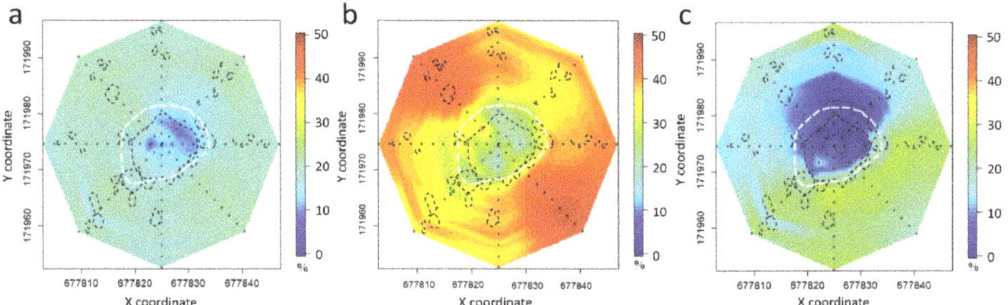

Figure 6. Spatial plots of exceedance rates (%) for VPD values with respect to 3.0 kPa from 24-h period of three measurement periods (coordinates refer to the Hungarian Unified National Projection System (m)); (**a**): May, (**b**): July, and (**c**): October. White dashed line: visual tree edge. Black dashed lines: the position and extent of individual shrubs and poplar sprouts.

Festuca vaginata (the only species in Group A) occurred both in the grasslands and in the grove along the transect in Figure 8. Except for a few grassland species that were found on both sides of the group of trees (Group B), there were three characteristic spatial groups of species: species occurring on the south-eastern side (C) or the north-western side (D) of the grassland and species occurring mainly or exclusively in the grove (E). According to the Zólyomi-indicator values, the plant species below the canopy had a notably higher moisture and lower temperature requirement than the species in the grassland. On the other hand, there was also a difference between the species of the two grassland sides.

Group C (species occurring on the south-eastern side of the grassland) had slightly lower moisture and higher temperature requirements than Group D (species occurring on the north-western side of the grassland).

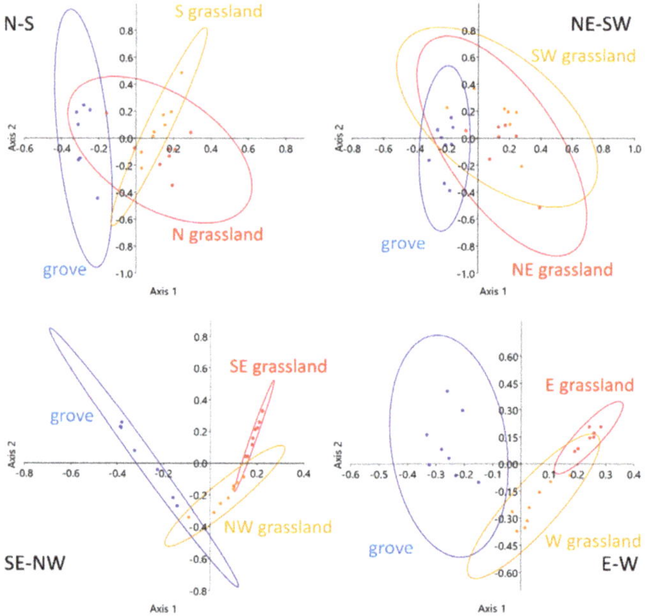

Figure 7. Principal coordinate analysis (PCoA) ordination of VPD quantiles in each transect in July 2018; 95% confidence ellipses are drawn according to vegetation patches.

Table 1. Species groups of the SE–NW transect ordered according to Figure 8. (A) ubiquitous species; (B) species occurring in the grassland; (C) species occurring on the south-eastern side of the grassland; (D) species occurring on the north-western side of the grassland; (E) species occurring in the grove; (F) low-frequency species not marked in Figure 8.

A	*Festuca vaginata*
B	*Cladonia foliacea* Huds.; *Tortula ruralis* (Hedw.) Gaertn., Meyer, & Scherb.; *Euphorbia seguieriana* Neck.; *Potentilla arenaria* Borkh.; *Artemisia campestris* L.
C	*Stipa borysthenica* Klokov ex Prokudin; *Dianthus serotinus* Waldst. et Kit.; *Fumana procumbens* Gren. et Godr.; *Bothriochloa ischaemum* (L.) Keng; *Poa bulbosa* L.; *Thymus pannonicus* All.; *Cladonia furcata* (Huds.) Schrad.; *Syrenia cana* (Piller et Millerp.) Neilr.
D	*Poa angustifolia* L.; *Gypsophila fastigiata* L.; *Cladonia magyarica* Vain. ex Gyeln.; *Carex liparicarpos* Gaudin; *Eryngium campestre* L.; *Asperula cynanchica* L.; *Salix rosmarinifolia* L.; *Centaurea arenaria* M. Bieb, ex Willd.; *Cynodon dactylon* (L.) Pers.; *Alyssum tortuosum* Willd.; *Thesium linophyllon* L.; *Equisetum ramosissimum* Desf.; *Tragopogon floccosus* Waldst. et Kit.; *Scirpoides holoschoenus* (L.) Soják; *Stipa capillata* L.; *Hieracium auriculoides* Láng
E	*Calamagrostis epigeios* (L.) Roth; *Crataegus monogyna* Jacq.; *Solidago virgaurea* L.; *Juniperus communis* L.; *Populus alba*; *Celtis occidentalis* L.
F	*Alkanna tinctoria* (L.) Tausch; *Asparagus officinalis* L.; *Berberis vulgaris* L.; *Chondrilla juncea* L.; *Cornus mas* L.; *Crepis rhoeadifolia* M. Bieb.; *Cynoglossum officinale* L.; *Euphorbia cyparissias* L.; *Galium mollugo* L. s. str.; *Helianthemum canum* (L.) Hornem.; *Leontodon autumnalis* L.; *Medicago minima* (L.) L.; *Polygonum arenarium* Waldst. et Kit.; *Taraxacum officinale* F.H.Wigg; *Teucrium chamaedris* L.; *Tragus racemosus* (L.) All.

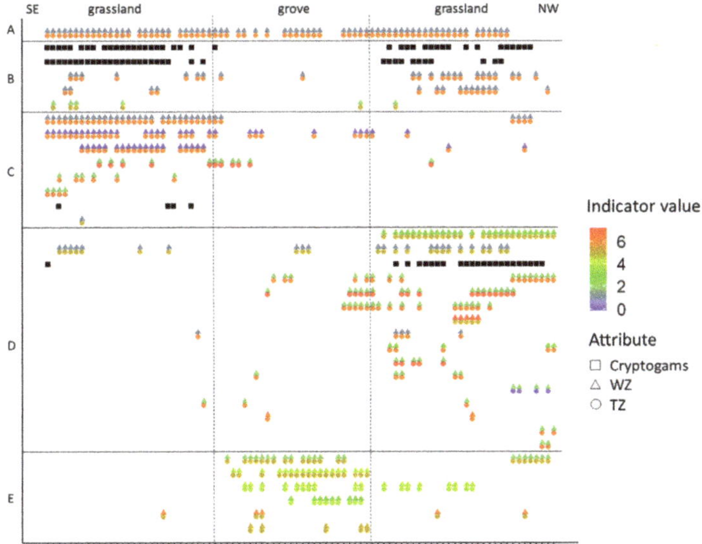

Figure 8. Distribution of the species along the SE–NW (Southeast–Northwest) transect in July 2018, together with the Zólyomi ecological indicator values, which quantify the temperature (TZ) and moisture (WZ) optimums for the plant species. Horizontal bars indicate coexisting species groups: (A) ubiquitous species; (B) species occurring in the grassland; (C) species occurring on the south-eastern side of the grassland; (D) species occurring on the north-western side of the grassland; (E) species occurring in the grove. List of species is shown in Table 1.

The ordination of the microcoenological data shows a clear difference between the three vegetation patches (Figure 9). The two grassland sides were notably separated from each other, and the grove just slightly overlapped with the NW grassland area. This slight overlap also applied to the VPD distribution (Figure 7, lower-left panel), indicating a more gradual transition toward the grove. The SE grassland side separated from the grove, similarly to the vegetation data, and the small overlap between the two grassland sides referred to the microclimatic similarities at the transect ends, which could also be observed in Figure 8. However, the two grassland sides showed remarkable microcoenological differences, and this difference was also manifested in the microclimatic conditions.

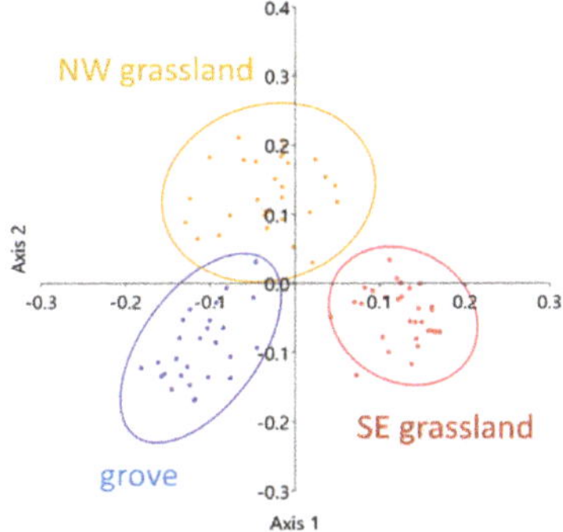

Figure 9. PCoA ordination of microcoenological data from the SE–NW transect. Jaccard index was used on summer measurements. Ellipses representing 95% confidence are drawn according to vegetation patch.

4. Discussion

Habitats sensitive to environmental changes, such as transition zones, have complex vegetation dynamics formed by the interaction of the biotic and abiotic factors of the two vegetation components [4]. This interaction is called an edge effect, which is manifested in the microclimatic and vegetation structural differences between the forest edge and the interior of the forest [5]. In the sandy forest-steppe habitat, the edge effect is caused by the different structural complexity of the surrounding grassland, with the outstandingly open, treeless, plain vegetation structure and low biomass [5,10,14].

To test some of the abiotic–biotic relationships for transition zones, we carried out microclimate measurements in a sandy forest-steppe habitat in Hungary. From the air temperature and air humidity measured, we calculated the vapour pressure deficit (VPD), which is an important limiting factor in plant growth, determining plant photosynthesis, because CO_2 uptake is strongly limited due to stomatal closure to prevent water loss [5,22,24].

We found that during a 24-h period, microclimatic differences between grassland, edge, and below-canopy areas can be determined based on the VPD (Figure 2). The range of the VPD was significantly lower within the grove than in the open area. These results are in good agreement with previous temperature data of several other edge studies [4,5,10,13]. In the edges, we also observed a special behaviour of the VPD in that the values were higher with a more abrupt rise in the warmer edge than in the open area, which can be explained by the heat-reflecting properties of the sunny side of the tree group. However, on the colder side of the transects, in line with the shadow effect, this elevated VPD does not occur at the edges and the deficit increases with distance from the tree group. This effect was more pronounced due to the lower altitude of the Sun (October) in the eastern, south-eastern, and southern borders but also in relatively cloudy weather conditions (May).

However, the extreme values, such as the maximum and the minimum, often occur only for short periods. We introduced a new method of applying the duration curve, showing the distribution of a variable over a measurement period [34,35]. Thus, the duration of values being over a threshold provides important information about the spatio-temporal VPD pattern. In this case, the exceedance rate with respect to the critical value of 1.2 kPa calculated from the 24-h VPD DCs showed a stress effect [22,24]. The duration curve method is presented in Figure 3 for the SE–NW transect, where the difference in the effect of the exposure was the most pronounced, in contrast to that for the N–S orientation, which is referred to as the coolest–warmest gradient in other studies [4,13,20]. For example, a study [4] of the edges of an oak-chestnut forest in the U.S. showed a difference of more than 5 °C between the south and north edges at the same measurement time. Another study in a Douglas-fir forest in the U.S. [13] also observed a remarkable difference between the south and north edges in short-wave radiation and air humidity during a diurnal period. In our study, in each season, at least 20% of the values were above the threshold, indicating semi-desert conditions in this habitat [9,10]. This new method also showed an increase in the variability of the exceedance rates (Figure 3), thus highlighting the importance of examining significant differences between the parts of the study site with a higher threshold.

We also found seasonal differences among the VPD distributions: the exceedance rate was much lower in spring and autumn than in the summer measurements (Figures 4 and 6). Spring and autumn did not differ significantly from each other due to the cloudy weather conditions in May. We also found differences in the VPD exceedance rate between opposite ends of the transects, which is a clear indication of the stronger stress effects in the warmer grassland areas during autumn. Choosing 3.0 kPa as the threshold value for the exceedance rate, the curves show a stronger moderating effect of the grove: a 0–10% duration of values being above 3.0 kPa in the colder grassland areas as opposed to a 25–30% duration in the warmer areas in the autumn. In the case of the above-3.0 kPa rate, this significant difference between the opposite areas showed that the distribution of the values above the threshold was notably varied within a seemingly homogeneous grassland (Figures 3 and 5). Thus, our first hypothesis that the VPD-modifying effect of the grove would gradually decrease in all directions away from the edge was only partially confirmed because this modifying effect can be detected even at a distance of 6–10 meters from the edge, which is slightly different from the results of the other edge studies. Considering the small size of the grove we examined, this effect was stronger than suggested in other studies [4,5,10,13]. Our second hypothesis, which says that the spatial microclimate patterns do not differ from season to season and only the intensity of the modifying effect changes should also be rejected since the seasonal differences were also manifest in the changes in the position of the edge effect.

In the edges of the grove, we found an interesting pattern in the VPD exceedance rate (Figure 5a,c). We observed an increasing tendency from west to south-east and then a remarkable decreasing trend from south to west. However, these trends did not appear in the case of the grassland end positions (Figure 5b,d). Thus, the cardinal and intercardinal directions clearly define the characteristics of stressful conditions in the edges of the forest patches. During the summer measurements, a sharp temperature difference developed at the southern, southeastern, and eastern edges at higher altitudes of the Sun, while no pronounced edge effect was detected in the north-eastern and south-western borders. Due to the exposure and the distance from the grove, the open areas were also influenced by other parameters such as the altitude and shadows of other groves and vegetation patches. Other studies [4,20] showed that due to the larger exposure to irradiation, the south edge should differ the most from the forest interior. In the U.S., oak-chestnut forests showed a difference of at least 5 °C between the south edge and the forest interior [4]. According to our measurements, this heat gradient did not develop in summer (Figure 6b), but it was visible in the SW–NE line in autumn (Figure 6c); thus, the southern areas were not persistently the warmest in each season. Above the 3.0 kPa threshold in October (Figure 5c), in the NW, N, and NE edges, the exceedance rate was 0%, which is a good indication of the shading effect due to the lower altitude of the Sun in October. Ordinations based on the distribution of the VPD values (Figure 7) also support the microclimate-modifying effect of the grove. This effect was present in all periods, but it was most pronounced during the summer measurements. We also found clear differences between the edges based on the cardinal and intercardinal directions with this analysis, which also partially refutes our first hypothesis stating that similar trends of VPD-modifying effect can be observed in all directions from the grove.

The patterns explored highlight the importance of fine-scale sampling and analysis. Owing to our microcoenological data, we also found spatial heterogeneity in plant species distributions (Figure 8) in the open sandy grassland plant associations (*Festucetum vaginatae*). Except for a few grassland species that were found on both sides of the group of trees, there were three characteristic spatial groups of species: species occurring on the south-eastern side or the north-western side of the grassland and species occurring in the grove. This indicated the microclimatic differences in the study site, mostly between the grove and the surrounding, opposite grassland areas. Some species within the association behaved as differential species and indicated well the microclimatic differences in a seemingly homogeneous grassland area with a small group of trees. This distribution of plant species is a good indication of the different environmental conditions underneath even such a small group of trees. In accordance with the spatial heterogeneity of plant species distributions in our study, previous studies also showed that the forest edge may have a higher species number than the surrounding open area or the forest interior; thus, the edge acts as a transition zone with higher diversity [5,10,12,14,42,43]. In our case, the shading effect of the tree group has significant influencing power, which can be observed in the surrounding grassland. Therefore, our third hypothesis that the coenological and indication structure of the herb layer is homogeneous in the open grassland must also be rejected.

5. Conclusions

This study has provided information concerning the effects of a small group of trees on the microclimate components of a forest-steppe habitat. We observed the main characteristics and distribution of the dominant physical parameters of the abiotic edge effect [14], including the air temperature and air humidity in this habitat. The vapour pressure deficit is a susceptible indicator of the environmental conditions for vegetation. We successfully applied the duration curve method on the VPD dataset. The VPD duration curves with thresholds of 1.2 kPa and 3.0 kPa were very informative in the spatio-temporal analysis, as the VPD exceedance rate could show the stress rate for the vegetation.

The cardinal and intercardinal directions as well as the altitude of the Sun have influences on the moderating and shading effects of the grove. With distance from the grove, the effects of other parameters increase significantly. We have also shown that not only the southern area could be consistently the warmest in a natural ecosystem.

We highlighted the importance of fine-scale sampling and analysis by demonstrating that the sample positions densely spaced in a continuous transect were able to reveal significant differences within small distances, which could be important in reassessing the structure of transition zones. In addition, they can uncover the patterns within seemingly homogeneous vegetation.

This knowledge is valuable for assessing the dynamics and spatio-temporal pattern of abiotic factors and physiognomy in this type of ecosystem, which is a natural transition zone in the temperate vegetation.

Supplementary Materials: The following are available online at http://www.mdpi.com/1999-4907/11/10/1078/s1. Table S1: Zólyomi's T and W scales (TZ, WZ)—relative ecological indicator values for the optimum temperature and humidity.

Author Contributions: Conceptualization, L.K. and J.B.; methodology, G.S. and L.K.; software, S.F.; sampling, G.S., B.G., and L.K.; validation, L.K.; formal analysis, G.S. and L.K.; writing—original draft preparation, G.S.; writing—review and editing, L.K., J.B., and S.F.; visualization, G.S., S.F., and L.K. All authors have read and agreed to the published version of the manuscript.

Funding: This research received no external funding.

Acknowledgments: We would like to thank the Institute of Ecology and Botany, Centre for Ecological Research, for use of the Fülöpháza research station. G.S., J.B., and S.F. acknowledge the support of the Higher Education Institutional Excellence Program (NKFIH-1159-6/2019) awarded by the Ministry of Human Capacities within the framework of water-related research of Szent István University.

Conflicts of Interest: The authors declare no conflict of interest.

References

1. Von Arx, G.; Dobbertin, M.; Rebetez, M. Spatio-temporal effects of forest canopy on understory microclimate in a long-term experiment in Switzerland. *Agric. For. Meteorol.* **2012**, *166*, 144–155. [CrossRef]
2. Bertrand, R.; Lenoir, J.; Piedallu, C.; Riofrı´o-Dillon, G.; de Ruffray, P.; Vidal, C.; Pierrat, J.C.; Ge´gout, J.C. Changes in plant community composition lag behind climate warming in lowland forests. *Nature* **2011**, *479*, 517–520. [CrossRef]
3. Cunningham, C.; Zimmermann, N.E.; Stoeckli, V.; Bugmann, H. Growth of Norway spruce (*Picea abies* L.) saplings in subalpine forests in Switzerland: Does spring climate matter? *For. Ecol. Manage.* **2006**, *228*, 19–32. [CrossRef]
4. Matlack, G.R. Microenvironment variation within and among deciduous forest edge sites in the eastern United State. *Biol. Conserv.* **1993**, *66*, 185–194. [CrossRef]
5. Young, A.; Mitchell, N. Microclimate and vegetation edge effects in a fragmented podocarp-broadleaf forest in New Zealand. *Biol. Conserv.* **1994**, *67*, 63–72. [CrossRef]
6. Baker, S.C.; Spies, T.A.; Wardlaw, T.J.; Balmer, J.; Franklin, J.F.; Jordan, G.J. The harvested side of edges: Effect of retained forests on the re-establishment of biodiversity in adjacent harvested areas. *For. Ecol. Manag.* **2013**, *302*, 107–121. [CrossRef]
7. Heithecker, T.D.; Halpern, C.B. Edge-related gradients in microclimate in forest aggregates following structural retention harvests in western Washington. *For. Ecol. Manag.* **2007**, *248*, 163–173. [CrossRef]
8. Kovács, B.; Tinya, F.; Ódor, P. Stand structural drivers of microclimate in mature temperate mixedforests. *Agric. For. Meteorol.* **2017**, *234*, 11–21. [CrossRef]
9. Bodrogközy, G. Die Vegetation der Weisspappel-Haine in dem Reservat Emlékerdő bei Szeged-Ásotthalom. *Acta Biol. Szeged.* **1957**, *3*, 127–140.

10. Erdős, L.; Tölgyesi, C.; Horzsea, M.; Tolnay, D.; Hurton, Á.; Schulcz, N.; Körmöczi, L.; Lengyel, A.; Bátori, Z. Habitat complexity of the Pannonian forest-steppe zone and its nature conservation implications. *Ecol. Complex.* **2014**, *17*, 107–118. [CrossRef]

11. Pongrácz, R.; Bartholy, J.; Kis, A. Estimation of future precipitation conditions for Hungary with special focus on dry periods. *Időjárás Q. J. Hung. Meteorol. Serv.* **2014**, *118*, 305–321.

12. Morecroft, M.D.; Taylor, M.E.; Oliver, H.R. Air and soil microclimates of deciduous woodland compared to an open site. *Agric. For. Meteorol.* **1998**, *90*, 141–156. [CrossRef]

13. Chen, J.; Franklin, J.F.; Spies, T.A. Contrasting microclimates among clearcut, edge, and interior old-growth Douglas-fir forest. *Agric. For. Meteorol.* **1993**, *63*, 219–237. [CrossRef]

14. Murcia, C. Edge effects in fragmented forests: Implications for conservation. *Trends Ecol. Evol.* **1995**, *10*, 58–62. [CrossRef]

15. Latif, Z.A.; Blackburn, G.A. The effects of gap size on some microclimate variables during late summer and autumn in a temperate broadleaved deciduous forest. *Int. J. Biometeorol.* **2010**, *54*, 119–129. [CrossRef]

16. Potter, C. Microclimate influences on vegetation water availability and net primary production in coastal ecosystems of Central California. *Landsc. Ecol.* **2014**, *29*, 677–687. [CrossRef]

17. Cuena-Lombraña, A.; Fois, M.; Fenu, G.; Cogoni, D.; Bacchetta, G. The impact of climatic variations on the reproductive success of *Gentiana lutea* L. in a Mediterranean mountain area. *Int. J. Biometeorol.* **2018**, *62*, 1283–1295. [CrossRef]

18. Hohnwald, S.; Indreica, A.; Walentowski, H.; Leuschner, C. Microclimatic Tipping Points at the Beech-Oak Ecotone in the Western Romanian Carpathians. *Forests* **2020**, *11*, 919. [CrossRef]

19. Lin, B.S.; Lin, Y.J. Cooling Effect of Shade Trees with Different Characteristics in a Subtropical Urban Park. *Hortscience* **2010**, *45*, 83–86. [CrossRef]

20. Chen, J.; Franklin, J.F.; Spies, T.A. Growing-season microclimatic gradients from clearcut edges into old-growth douglas-fir forests. *Ecol. Appl.* **1995**, *5*, 74–86. [CrossRef]

21. Bolton, D. The computation of equivalent potential temperature. *Mon. Weather Rev.* **1980**, *108*, 1046–1053. [CrossRef]

22. Novick, K.; Ficklin, D.; Stoy, P.; Williams, C.A.; Oishi, A.C.; Papuga, S.A.; Blanken, P.D.; Noormets, A.; Sulman, B.N.; Scott, R.; et al. The increasing importance of atmospheric demand for ecosystem water and carbon fluxes. *Nat. Clim. Chang.* **2016**, *6*, 1023–1027. [CrossRef]

23. Spittlehouse, D.L.; Adams, R.S.; Winkler, R.D. *Forest, Edge and Opening Microclimate at Sicamous Creek*; Research Report 24; Ministry of Forests Research Branch: Victoria, BC, Canada, 2004.

24. Shamshiri, R.R.; Jones, J.W.; Thorp, K.R.; Ahmad, D.; Man, H.C.; Taheri, S. Review of optimum temperature, humidity, and vapour pressure deficit for microclimate evaluation and control in greenhouse cultivation of tomato: A review. *Int. Agrophysics* **2018**, *32*, 287–302. [CrossRef]

25. Shamshiri, R.R.; Kalantari, F.; Ting, K.C.; Thorp, K.R.; Hameed, I.A.; Weltzien, C.; Ahmad, D.; Shad, Z.M. Advances in greenhouse automation and controlled environment agriculture: A transition to plant factories and urban agriculture. *Int. J. Agricul. Biol. Eng.* **2018**, *11*, 1–22. [CrossRef]

26. Geiger, R.; Aron, R.H.; Todhunter, P. *The Climate Near the Ground*; Rowman & Littlefield: Lanham, MD, USA, 2009; 623p. [CrossRef]

27. Magnago, L.F.S.; Rocha, M.F.; Meyer, L.; Martins, S.V.; Meira-Neto, J.A.A. Microclimatic conditions at forest edges have significant impacts on vegetation structure in large Atlantic forest fragments. *Biodivers. Conserv.* **2015**, *24*, 2305–2318. [CrossRef]

28. Connell, J.H.; Slatyer, R.O. Mechanisms of succession in natural communities and their role in community stability and organization. *Am. Nat.* **1977**, *111*, 1119–1144. [CrossRef]

29. Kollmann, J. Regeneration window for fleshy-fruited plants during scrub development on abandoned grassland. *Ecoscience* **1995**, *2*, 213–222. [CrossRef]

30. Wiström, B.; Nielsen, A.B. Effects of planting design on planted seedlings and spontaneous vegetation 16 years after establishment of forest edges. *New For.* **2014**, *45*, 97–117. [CrossRef]

31. Körmöczi, L.; Bátori, Z.; Erdős, L.; Tölgyesi, C.; Zalatnai, M.; Varró, C. The role of randomization tests in vegetation boundary detection with moving split-window analysis. *J. Veg. Sci.* **2016**, *27*, 1288–1296. [CrossRef]

32. Erdős, L.; Ambarlı, D.; Anenkhonov, O.A.; Bátori, Z.; Cserhalmi, D.; Kiss, M.; Kröel-Dulay, G.; Liu, H.; Magnes, M.; Molnár, Z.; et al. The edge of two worlds: A new review and synthesis on Eurasian forest-steppes. *Appl. Veg. Sci.* **2018**, *21*, 345–362. [CrossRef]

33. FLÓRA Adatbázis 1.2 Taxonlista és Attribútum-Állomány (FLORA Database 1.2 Taxon-List and Attribute Values). Available online: https://okologia.mta.hu/node/2448 (accessed on 12 September 2020).

34. Vogel, R.M.; Fennessey, N.M. Flow-duration curves. I: New interpretation and confidence intervals. *J. Water Res. Plan. Manag.* **1994**, *120*, 485–504. [CrossRef]

35. Lane, P.N.J.; Best, A.E.; Hickel, K.; Zhang, L. The response of flow duration curves to afforestation. *J. Hydrol.* **2005**, *310*, 253–265. [CrossRef]

36. Verma, R.K.; Murthy, S.; Verma, S.; Mishra, S.K. Design flow duration curves for environmental flows estimation in Damodar River Basin, India. *Appl. Water Sci.* **2017**, *7*, 1283–1293. [CrossRef]

37. R Core Team. *R: A Language and Environment for Statistical Computing*; R Foundation for Statistical Computing: Vienna, Austria, 2017.

38. Akima, H.; Gebhardt, A.; Petzold, T.; Maechler, M. Interpolation of Irregularly and Regularly Spaced Data. R Package Version 0.6–2. 2016. Available online: https://cran.r-project.org/web/packages/akima/akima.pdf (accessed on 10 March 2020).

39. Oksanen, J.; Blanchet, F.G.; Friendly, M.; Kindt, R.; Legendre, P.; McGlinn, D.; Minchin, P.R.; O'Hara, R.B.; Wagner, H. Vegan: Community Ecology Package. R Package Version 2.5–6. 2019. Available online: http://vegan.r-forge.r-project.org/ (accessed on 18 March 2020).

40. Wickham, H. *ggplot2: Elegant Graphics for Data Analysis*; Springer: New York, NY, USA, 2016.

41. Hall, C.A.; Meyer, W.W. Optimal error bounds for cubic spline interpolation. *J. Approx. Theory* **1976**, *16*, 105–122. [CrossRef]

42. Ries, L.; Fletcher, R.J.; Battin, J.; Sisk, T.D. Ecological responses to habitat edges: Mechanisms, models, and variability explained. *Annu. Rev. Ecol. Evol. Syst.* **2004**, *35*, 491–522. [CrossRef]

43. Erdős, L.; Török, P.; Szitár, K.; Bátori, Z.; Tölgyesi, C.; Kiss, P.J.; Bede-Fazekas, Á.; Kröel-Dulay, G. Beyond the Forest-Grassland Dichotomy: The Gradient-Like Organization of Habitats in Forest-Steppes. *Front. Plant. Sci.* **2020**, *11*, 236. [CrossRef]

Article

Climatic, Edaphic and Biotic Controls over Soil δ^{13}C and δ^{15}N in Temperate Grasslands

Xing Zhao [1], Xingliang Xu [2,3,*], Fang Wang [1], Isabel Greenberg [4], Min Liu [1,2], Rongxiao Che [5], Li Zhang [6] and Xiaoyong Cui [1,3,7,*]

[1] University of Chinese Academy of Sciences, Beijing 100049, China; zhaoxing@ucas.edu.cn (X.Z.); wangfang13@mails.ucas.edu.cn (F.W.); lium.16s@igsnrr.ac.cn (M.L.)

[2] Key Laboratory of Ecosystem Network Observation and Modeling, Institute of Geographic Sciences and Natural Resources Research, Chinese Academy of Sciences, 11A, Datun Road, Chaoyang District, Beijing 100101, China

[3] CAS Center for Excellence in Tibetan Plateau Earth Sciences, Chinese Academy of Sciences, Beijing 100101, China

[4] Department of Environmental Chemistry, University of Kassel, Nordbahnhofstrasse 1a, 37213 Witzenhausen, Germany; isabel.greenberg@uni-kassel.de

[5] Institute of International Rivers and Eco-Security, Yunnan University, Kunming 650500, China; cherongxiao@ynu.edu.cn

[6] Ministry of Education Key Laboratory for Biodiversity Science and Ecological Engineering, Coastal Ecosystems Research Station of the Yangtze River Estuary, Shanghai Institute of Eco-Chongming, School of Life Sciences, Fudan University, 2005 Songhu Road, Shanghai 200438, China; zhangl1222@outlook.com

[7] Yanshan Eco-Environmental Observatory, Chinese Academy of Sciences, Beijing 101408, China

* Correspondence: xuxingl@hotmail.com (X.X.); cuixy@ucas.edu.cn (X.C.)

Received: 17 March 2020; Accepted: 8 April 2020; Published: 10 April 2020

Abstract: Soils δ^{13}C and δ^{15}N are now regarded as useful indicators of nitrogen (N) status and dynamics of soil organic carbon (SOC). Numerous studies have explored the effects of various factors on soils δ^{13}C and δ^{15}N in terrestrial ecosystems on different scales, but it remains unclear how co-varying climatic, edaphic and biotic factors independently contribute to the variation in soil δ^{13}C and δ^{15}N in temperate grasslands on a large scale. To answer the above question, a large-scale soil collection was carried out along a vegetation transect across the temperate grasslands of Inner Mongolia. We found that mean annual precipitation (MAP) and mean annual temperature (MAT) do not correlate with soil δ^{15}N along the transect, while soil δ^{13}C linearly decreased with MAP and MAT. Soil δ^{15}N logarithmically increased with concentrations of SOC, total N and total P. By comparison, soil δ^{13}C linearly decreased with SOC, total N and total P. Soil δ^{15}N logarithmically increased with microbial biomass C and microbial biomass N, while soil δ^{13}C linearly decreased with microbial biomass C and microbial biomass N. Plant belowground biomass linearly increased with soil δ^{15}N but decreased with soil δ^{13}C. Soil δ^{15}N decreased with soil δ^{13}C along the transect. Multiple linear regressions showed that biotic and edaphic factors such as microbial biomass C and total N exert more effect on soil δ^{15}N, whereas climatic and edaphic factors such as MAT and total P have more impact on soil δ^{13}C. These findings show that soil C and N cycles in temperate grasslands are, to some extent, decoupled and dominantly controlled by different factors. Further investigations should focus on those ecological processes leading to decoupling of C and N cycles in temperate grassland soils.

Keywords: carbon cycling; natural stable isotope abundance; nitrogen cycling; soil organic matter; temperate grassland

---

1. Introduction

Soil organic matter (SOM) consists of a heterogeneous mixture of substances in various stages of decay, mainly including plant and animal residues, microbial necromass, and new substances synthesized and released by microbes into the soil [1,2]. The global SOM pool in the surface meter stores approximately 1500 Pg carbon (C) [3] and 95 Pg nitrogen (N) [4] as well as other essential elements for plants and microbes. Therefore, SOM is critical for soil quality and ecosystem dynamics [5,6]. At the same time, SOM plays an important role in global climate change because soils could act as a potential sink for C [7,8]. Therefore, a large number of studies have investigated the effects of various climatic, edaphic and biotic factors on the dynamics of C and N in soils to better understand their turnover and SOM destabilization as well as their role in climate change [9–13].

It is difficult to explore the dynamics of SOM by direct measurement of the change in C and N stocks due to their large size [14]. With the rapid development of isotope ratio mass spectrometry [15], the analysis of stable isotope composition of soil C (δ^{13}C) and N (δ^{15}N) has become a powerful tool to explore the stability and dynamics of SOM [12,16–18] and soil development [12,19]. Soils δ^{13}C and δ^{15}N are ideally suited to provide wider insights into C and N cycles in soil ecosystems because they are primarily based on either an isotopic fractionation during microbial degradation and transformation (e.g., ammonification, nitrification and denitrification) or the preferential decomposition of the substrates depleted in ^{13}C and ^{15}N [20]. Generally, older and more microbially-processed SOM is enriched in ^{13}C and ^{15}N compared to less-decomposed substrates [18,21].

Additionally, variation in δ^{13}C and δ^{15}N content of SOM in natural ecosystems is largely controlled by the input of new plant residues and overall isotopic fractionation during microbial decomposition [18]. The signature of δ^{13}C and δ^{15}N in SOM is closely related to vegetation changes and microbial decomposition as well as anthropogenic N input [22–24]. Moreover, climatic and edaphic factors, including temperature, precipitation, pH, and contents of soil C, N and phosphorus (P) as well as soil texture, greatly impact δ^{13}C and δ^{15}N content of SOM [13,25,26]. As a result, the signature of ^{13}C and ^{15}N in SOM can be used as a valid proxy for SOM dynamics and provide integrated information about the ecosystem N cycling [9,27–32].

To understand the factors controlling δ^{13}C and δ^{15}N in SOM, numerous studies have investigated the patterns of soil δ^{13}C and δ^{15}N on regional and global scales [9–11,14,31,33,34]. It has been shown that climate controls forest soil δ^{13}C in the southern Appalachian Mountains [13]. Climate can likewise have an effect on soil δ^{15}N, with values increasing in response to rain events, which enhance the processes that cause the loss of N but discriminate against ^{15}N loss [35]. Further evidence shows that aridity can nonlinearly alter soil δ^{15}N values in arid and semi-arid grasslands [34]. Consequently, soil δ^{15}N values along precipitation gradients can reflect the pattern of N losses relative to turnover [36–38]. On a global scale, soil δ^{15}N converges across climate and latitudinal gradients [11]. In addition to climatic factors, substrate age, soil texture and litter input as well as land-use change also can affect soil δ^{13}C and δ^{15}N [25,31,39,40]. Nonetheless, the controls on C and N isotope ratios in soil still remain unclear [27].

Grasslands are an interesting ecosystem to study in this context because they store large amounts of C and N in soil [41,42] and have great potential to affect CO_2 concentrations in the atmosphere. Additionally, grasslands are widely distributed over the world and account for 26% of the ice-free land [43]. As a result, grassland soils play an important role in the context of global climate change and regulate biogeochemical cycles [44]. Among the various types, temperate grasslands are widely distributed across the Eurasian continent and form the Eurasian steppe [45]. Recent studies of temperate grasslands showed that climatic variables control approximately 50% of the variation in soil δ^{15}N along an east–west transect in Northern China. Soil δ^{15}N was found to decrease with increasing mean annual precipitation (MAP) and mean annual temperature (MAT) [10]. Further studies demonstrated that the aridity can nonlinearly alter soil δ^{15}N values [34]. Nonetheless, it remains unclear how co-varying climatic, edaphic and biotic factors control soil δ^{13}C and δ^{15}N in such temperate grasslands. We hypothesize that distinct factors control the soil δ^{13}C versus δ^{15}N signature: (1) biotic factors such as microbial biomass C (MBC) and N (MBN) as well as plant belowground biomass could exert more impact on soil ^{15}N than edaphic and climatic factors since ^{15}N fractionation is largely controlled by biological processes [11]; and (2) climatic and edaphic factors have more effects on soil δ^{13}C than biological factors because water can strongly affect ^{13}C in plant tissues [46]. To test the above hypotheses, we collected soil and plant samples from temperate meadow steppes, temperate steppes and temperate deserts along a vegetation transect in Inner Mongolia.

2. Materials and Methods

2.1. Study Sites

This study was conducted along a 1280 km transect across Inner Mongolia from west to east in northern China (Table 1). The longitude of the transect ranged from 107°15′ to 122°17′ and the latitude ranged from 38°44′ to 50°12′. The region was characterized predominantly by an arid and semi-arid continental climate. MAP ranged from 154 to 517 mm and MAT ranged from 1 to 4 °C. The MAP and MAT of each sampling site were calculated from the NMIC (China National Meteorological Information Center). The main vegetation types across this transect were temperate meadow steppe, temperate steppe, and temperate desert. All the soils were classified as chestnut soil, corresponding to Calcicorthic Aridisol according to USDA Soil Taxonomy [47].

In 2014, a field campaign was conducted to collect soil and plant samples along this transect. In total, 22 sampling sites including six temperate meadow steppes, nine temperate steppes and seven temperate deserts were selected. At each site, four plots (1 m × 1 m) were randomly selected for soil and plant sampling. The aboveground plant parts were harvested for the estimation of aboveground net primary production (ANPP). Additionally, five soil cores (2.5 cm diameter, 5 cm depth) were collected randomly using a soil corer from each plot and mixed thoroughly as one composite sample. Living roots were carefully collected from the soil and washed by water and then dried for the estimation of belowground biomass. Soil samples were sieved through a 2.0 mm sieve and then separated into two parts: one was stored in a plastic bag and frozen at −20 °C for measurement of soil moisture, MBC and MBN, and the other was air-dried for measurements under natural conditions.

2.2. Analysis of Soil δ^{13}C and δ^{15}N

Dried soil samples were ground into powder using a ball mill (Retsch MM2; Retsch, Haan, Germany). Approximately 1 g soil was put into 5 mL centrifuge tube. To remove carbonate, 3 mL of 0.5 M HCl was added to the tubes overnight. Afterwards, samples were freeze-dried and washed with H_2O until a pH of 7.0 was reached. The soil was weighed into tin capsules for analysis of total N (N_t), soil organic C (SOC), δ^{13}C and δ^{15}N by continuous flow gas isotope ratio mass spectrometry (isoprime precisION, Elementar, Germany). The isotope results of soil C or N were calculated as follows: δ^{13}C/δ^{15}N (‰) = (R_{sample}/$R_{standard}$ − 1) * 1000, where R_{sample} and $R_{standard}$ are the ratios of ^{13}C/^{12}C or ^{15}N/^{14}N in the sample and standard. The standards for δ^{13}C and δ^{15}N are Pee Dee Belemnite and atmospheric molecular N, respectively. The standard deviation of repeated measurements of laboratory standards was ±0.15‰ for these isotope analyses.

2.3. Analysis of Soil Properties and Microbial Biomass

Total phosphorus (P_t) in the soil was measured using optical emission spectrometry (Optima 5300DV; PerkinElmer, Shelton, USA) after nitric-perchloric acid digestion [48,49]. Soil pH was measured by a dry soil-water ratio of 1:2. Soil MBC and MBN were determined using the chloroform fumigation–extraction method [50,51]. Briefly, 10 g fresh soil was extracted with 24 mL of 0.5 M K_2SO_4. An additional 10 g soil was fumigated with ethanol-free chloroform for 24 h and then extracted again in the same manner. All extracts were shaken for 1 h and filtered through 5895 paper. Total organic C and N concentrations in the K_2SO_4 extracts were measured with a Dimatec-100 TOC/TIC analyzer (Liqui TOCII, Elementar, Germany).

Table 1. Detailed information about climatic, edaphic and biotic factors at 22 sampling sites along the vegetation transect across Inner Mongolian temperate grasslands.

Site	Type	Latitude	Longitude	Altitude (m)	MAP (mm)	MAT (°C)	g kg^{-1}			mg kg^{-1}			Biomass (g m^{-2})		‰	
							SOC	Total N	Total P	MBC	MBN	soil pH	Aboveground	Belowground	δ^{13}C	δ^{15}N
1	Temperate meadow steppe	50°12′	119°43′	550	339.1	3.0	36.10 ± 0.79	3.36 ± 0.08	0.9	422.9 ± 89.0	135.7 ± 29.7	6.9	26.9 ± 3.4	785.6 ± 64.8	-25.3 ± 0.1	6.3 ± 0.6
2	Temperate meadow steppe	49°12′	120°22′	590	394.9	3.5	31.30 ± 1.16	2.82 ± 0.13	0.5	506.2 ± 56.8	244.5 ± 115.9	7.3	55.7 ± 15.5	388.2 ± 20.2	-26.8 ± 0.1	5.6 ± 0.2
3	Temperate meadow steppe	49°17′	119°56′	590	337.7	3.2	29.77 ± 0.12	2.76 ± 0.11	0.5	467.8 ± 41.7	154.4 ± 21.1	—	37.4 ± 22.9	537.7 ± 92.9	-25.1 ± 0.2	5.3 ± 0.4
4	Temperate meadow steppe	49°9′	119°50′	500	337.7	3.3	34.64 ± 0.89	2.96 ± 0.09	0.6	384.3 ± 52.9	145.7 ± 20.5	7.1	42.6 ± 52.0	287.3 ± 97.1	-25.7 ± 0.0	5.9 ± 0.2
5	Temperate meadow steppe	49°33′″	117°19′	683	281.7	2.5	17.84 ± 1.55	1.88 ± 0.15	0.5	353.4 ± 47.7	182.5 ± 51.5	8.0	146.6 ± 42.0	439.6 ± 117.8	-24.2 ± 0.3	3.8 ± 0.3
6	Temperate meadow steppe	48°22′	122°17′	450	517.5	4.4	53.36 ± 0.66	4.69 ± 0.10	0.8	660.7 ± 50.2	392.4 ± 63.7	7.2	66.7 ± 14.0	680.7 ± 170.6	-27.1 ± 0.1	3.8 ± 0.6
7	Temperate steppe	43°96′	115°86′	1000	282.6	2.4	12.60 ± 0.32	1.37 ± 0.03	0.4	249.4 ± 57.3	105.8 ± 14.3	7.5	125.9 ± 63.7	268.4 ± 183.3	-23.0 ± 0.4	3.9 ± 0.6
8	Temperate steppe	43°52′	119°22′	644	349.5	3.1	19.57 ± 0.99	1.90 ± 0.11	0.4	273.4 ± 19.6	95.4 ± 14.0	7.3	20.0 ± 0.9	147.6 ± 72.0	-22.7 ± 0.4	2.0 ± 0.5
9	Temperate steppe	43°15′	118°09′	749	377.8	3.0	2.48 ± 0.45	0.27 ± 0.04	0.2	95.9 ± 63.0	68.1 ± 50.4	7.3	70.0 ± 16.0	215.3 ± 168.3	-23.8 ± 0.6	-3.5 ± 3.7
10	Temperate steppe	43°15′	117°11′	1296	399.1	3.3	18.06 ± 1.30	1.85 ± 0.11	0.3	333.7 ± 55.2	180.1 ± 21.7	7.2	62.2 ± 12.5	386.5 ± 71.6	-24.6 ± 0.6	2.1 ± 0.7
11	Temperate steppe	43°43′	112°50′	993	158.8	1.5	3.21 ± 0.19	0.36 ± 0.03	0.5	147.7 ± 10.5	65.4 ± 16.4	8.1	25.4 ± 4.0	143.0 ± 132.0	-21.4 ± 0.7	0.0 ± 2.1
12	Temperate steppe	43°12′	116°09′	1298	335.7	2.8	7.64 ± 0.54	0.72 ± 0.03	0.3	189.4 ± 13.6	79.5 ± 10.6	7.8	31.8 ± 5.1	372.0 ± 62.2	-24.8 ± 0.6	2.6 ± 1.3
13	Temperate steppe	43°19′	119°35′	453	349.5	3.0	6.12 ± 0.09	0.73 ± 0.03	0.1	138.5 ± 35.0	116.5 ± 24.8	8.1	11.0 ± 2.3	345.8 ± 196.8	-21.2 ± 0.3	0.9 ± 0.9
14	Temperate steppe	42°32′	118°53′	794	408.3	3.1	9.23 ± 0.91	1.02 ± 0.09	0.2	157.1 ± 40.1	53.2 ± 12.0	7.9	77.2 ± 46.3	407.1 ± 206.4	-22.9 ± 0.6	1.3 ± 0.7
15	Temperate steppe	41°20′	112°51′	1760	373.7	2.6	10.21 ± 0.50	1.07 ± 0.04	0.7	218.9 ± 54.7	106.7 ± 20.1	8.1	35.5 ± 9.1	324.1 ± 38.5	-24.6 ± 0.3	5.0 ± 1.4
16	Temperate desert	43°21′	111°52′	960	154.5	1.2	1.78 ± 0.12	0.20 ± 0.02	0.3	94.0 ± 26.3	55.4 ± 26.7	8.1	7.8 ± 3.5	78.0 ± 37.9	-22.6 ± 0.7	-1.7 ± 2.4
17	Temperate desert	42°56′	110°50′	1071	200.3	1.3	4.03 ± 0.40	0.55 ± 0.05	0.4	149.4 ± 22.5	71.2 ± 13.5	8.1	6.5 ± 5.7	106.4 ± 95.5	-21.5 ± 0.8	1.9 ± 1.8
18	Temperate desert	42°25′	109°49′	1158	188.2	1.4	3.95 ± 0.58	0.56 ± 0.07	0.3	158.1 ± 10.8	102.0 ± 52.3	8.3	19.6 ± 8.8	118.1 ± 64.8	-22.3 ± 0.4	3.1 ± 1.1
19	Temperate desert	41°54′	108°42′	1533	180.1	1.5	5.65 ± 0.24	0.67 ± 0.04	0.4	166.2 ± 15.4	84.1 ± 13.4	8.0	11.6 ± 3.2	179.0 ± 187.0	-23.6 ± 0.3	3.4 ± 1.5
20	Temperate desert	40°01′	110°03′	1339	347.1	2.8	2.88 ± 0.10	0.26 ± 0.02	0.4	76.2 ± 13.8	46.5 ± 16.0	8.3	31.5 ± 9.2	205.8 ± 92.0	-23.9 ± 0.6	0.4 ± 1.7
21	Temperate desert	39°14′	107°16′	1281	205.8	2.0	5.62 ± 0.55	0.51 ± 0.04	0.3	155.7 ± 14.3	76.8 ± 18.2	8.1	22.6 ± 5.4	58.5 ± 27.3	-22.7 ± 0.2	2.6 ± 0.9
22	Temperate desert	38°44′	107°45′	1345	267.6	2.4	1.97 ± 0.16	0.17 ± 0.02	0.4	71.8 ± 22.6	46.5 ± 24.8	8.7	87.4 ± 85.1	50.9 ± 44.9	-24.1 ± 0.6	0.3 ± 3.4

The values are means ± standard errors of 4 replicates. MAP = mean annual precipitation, MAT = mean annual temperature, SOC = soil organic carbon, N = nitrogen, P = phosphorous, MBC = microbial biomass C, MBN = microbial biomass N.

2.4. Calculations and Statistics

MBC and MBN were calculated as the difference between the total C or total N content in fumigated and non-fumigated soils, divided by a k_{EC} factor of 0.45 [52] and a k_{EN} factor of 0.54, respectively [50,51].

The standard errors of means are presented in figures and tables as a variability parameter. The normality of soil $\delta^{15}N$ and $\delta^{13}C$, as well as, other edaphic and biological data were tested. A one-way analysis of variance was performed with SPSS 21.0 (SPSS Inc., Chicago, IL, USA) to evaluate the effects of grassland type on soil $\delta^{15}N$ and $\delta^{13}C$ values. Correlations between soil $\delta^{15}N$ and $\delta^{13}C$ and climatic (MAP, MAT), edaphic (SOC, Nt, Pt) and biotic factors (MBC, MBN and belowground biomass) were analyzed with SPSS 21.0 (SPSS Inc., Chicago, IL, USA). To identify how all the factors affect soils $\delta^{15}N$ and $\delta^{13}C$, we conducted a data analysis in two steps using R version 3.5.2 (R Development Core Team 2019). The first step was to generate a series of all possible multiple linear models based on the information-theoretic method. To avoid overfitting our models, a Pearson correlation test was conducted to identify and remove highly correlated factors ($r > 0.6$ or < -0.6, Table 2) within one model. The second step was to calculate estimates and the relative importance of predictors considering changes to the models' Akaike's information criterion (AIC) changes of less than 2 (model.avg function in MuMIn package) with the model averaging method [53]. Information-theoretic AIC corrected for small samples sizes (AICc), ΔAIC (difference between AICc of one model and the model with the lowest AICc), and AICc weight (wAICc) were calculated for model ranking. All differences were tested for significance ($P < 0.05$).

Table 2. Pearson's correlation matrix for raw input variables in explaining change in soil ^{15}N and ^{13}C along the vegetation transect across Inner Mongolian temperate grasslands. The asterisks indicate a significant relationship between variables at $P < 0.05$.

	MAP	MAT	SOC	Total N	Total P	pH	MBC	MBN	BB
MAP	1.000								
MAT	0.959 *	1.000							
SOC	0.629 *	0.742 *	1.000						
Total N	0.627 *	0.732 *	0.996 *	1.000					
Total P	0.273	0.320	0.721 *	0.716 *	1.000				
pH	−0.584 *	−0.676 *	−0.780 *	−0.788	−0.402	1.000			
MBC	0.580 *	0.688 *	0.966 *	0.973 *	0.677 *	−0.742 *	1.000		
MBN	0.626 *	0.675 *	0.867 *	0.870 *	0.575 *	−0.579 *	0.919 *	1.000	
BB	0.677 *	0.697 *	0.781 *	0.802 *	0.602 *	−0.670 *	0.771 *	0.677 *	1.000

MAP = mean annual precipitation, MAT = mean annual temperature, SOC = soil organic carbon, N = nitrogen, P = phosphorous, MBC = microbial biomass C, MBN = microbial biomass N, BB = belowground biomass.

3. Results

3.1. Climatic, Edaphic and Biotic Factors along the Transect

Along the transect, MAT ranged from 1.2 to 4.5 °C with a mean of 2.6 °C and MAP ranged from 155 to 518 mm with a mean of 308 mm (Table 1). SOC across 22 sites varied from 1.97 to 53.36 g kg^{-1}, with the lowest value in temperate desert and the highest value in the temperate meadow steppe (Table 1). N_t ranged from 0.17 to 4.69 g kg^{-1}, while P_t varied from 0.1 to 0.9 g kg^{-1}. Soil C:N ratios were between 9.0 and 13.0. Soil pH varied from 7.1 to 8.3 (Table 1). MBC ranged from 71.8 to 660.5 mg kg^{-1} and MBN varied from 46.5 to 392.4 mg kg^{-1} (Table 1). Aboveground biomass was in the range of 11.6 to 146.6 g dry weight (d.w.) m^{-2}, while belowground biomass varied from 50.9 to 785.6 g d.w. m^{-2} (Table 1). All observed parameters decreased from temperate meadow steppe to temperate steppe to temperate desert along the transect.

3.2. Soil $\delta^{15}N$ and $\delta^{13}C$

Soil $\delta^{15}N$ cross 22 sites along this transect varied from −3.53‰ to 5.88‰ (Table 1). All seven temperate meadow steppe sites had positive $\delta^{15}N$ values while some temperate steppe and temperate desert sites had negative $\delta^{15}N$ values. Soil $\delta^{15}N$ in temperate meadow steppes (5.12‰) were higher than those in temperate steppes and temperate deserts ($P < 0.05$). By comparison, soil $\delta^{13}C$ cross the grassland transect ranged from −27.1‰ to −21.2‰ (Table 1). Soil $\delta^{13}C$ values in temperate meadow steppes (−25.7‰) were lower than those in temperate steppes and temperate deserts ($P < 0.05$).

3.3. Correlation Climatic, Edaphic and Biotic Factors with Soil $\delta^{15}N$ and $\delta^{13}C$

Along the transect, soil $\delta^{15}N$ were not correlated with MAP (Figure 1a) and MAT (Figure 1b). However, soil $\delta^{13}C$ linearly decreased with MAP ($R^2 = 0.43$, $P < 0.001$, Figure 1c) and MAT ($R^2 = 0.52$, $P < 0.001$, Figure 1d).

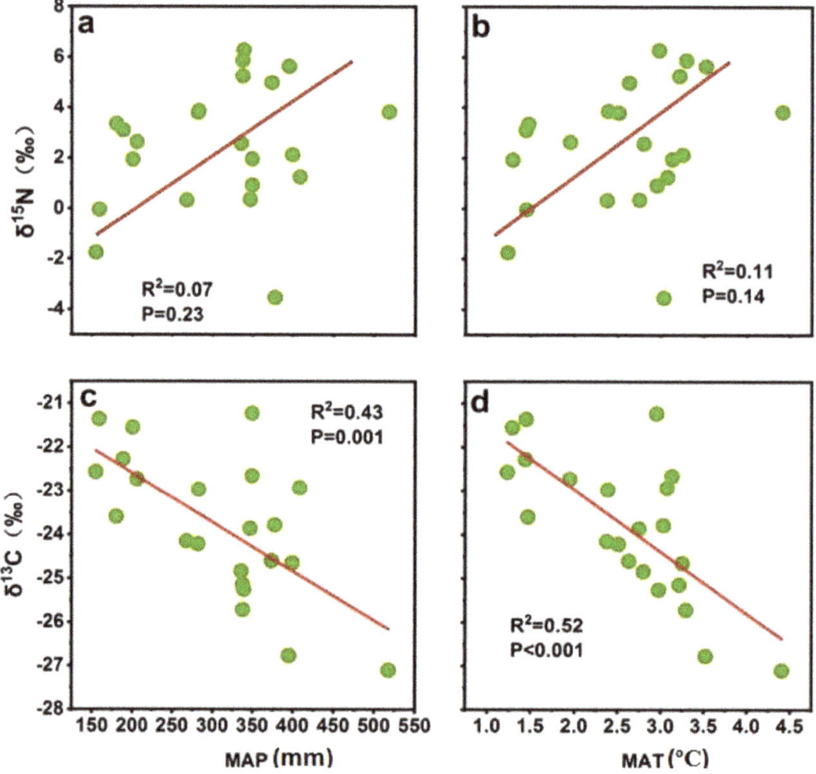

Figure 1. Relationships between soil $\delta^{15}N$ and $\delta^{13}C$ in the upper 5 cm with mean annual precipitation (MAP, **a**, **c**) and mean annual temperature (MAT, **b**, **d**) at 22 sites along the vegetation transect across Inner Mongolian temperate grasslands.

Soil $\delta^{15}N$ logarithmically increased with concentrations of SOC ($R^2 = 0.71$, $P < 0.001$, Figure 2a), N_t ($R^2 = 0.71$, $P < 0.001$, Figure 2b) and P_t ($R^2 = 0.51$, $P < 0.001$, Figure 2c). By comparison, soil $\delta^{13}C$ linearly decreased with SOC ($R^2 = 0.58$, $P < 0.001$, Figure 2d), N_t ($R^2 = 0.53$, $P < 0.001$, Figure 2e) and P_t ($R^2 = 0.37$, $P < 0.003$, Figure 2f).

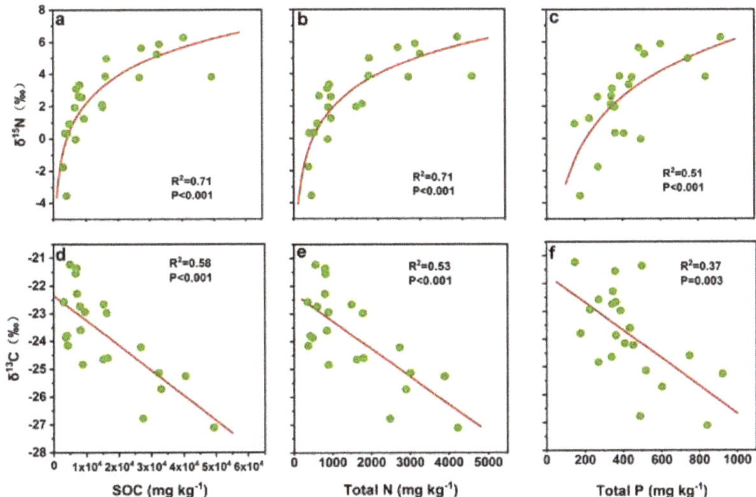

Figure 2. Relationships between soil δ^{15}N and δ^{13}C in the upper 5 cm with soil organic carbon (SOC, **a**, **d**), total nitrogen (N, **b**, **e**) and total phosphorus (P, **c**, **f**) at 22 sites along the vegetation transect across Inner Mongolian temperate grasslands.

Similarly to SOC and N$_\text{t}$, soil δ^{15}N logarithmically increased with concentrations of MBC ($R^2 = 0.63$, $P < 0.001$, Figure 3a) and MBN ($R^2 = 0.48$, $P < 0.001$, Figure 3b). Soil δ^{13}C linearly decreased with MBC ($R^2 = 0.58$, $P < 0.001$, Figure 3c) and MBN ($R^2 = 0.58$, $P < 0.001$, Figure 3d). Plant belowground biomass linearly increased with Soil δ^{15}N ($R^2 = 0.37$, $P = 0.004$, Figure 3e), but decreased with soil δ^{13}C ($R^2 = 0.41$, $P = 0.002$, Figure 3f). Soil δ^{15}N decreased with soil δ^{13}C along the transect ($R^2 = 0.30$, $P = 0.008$, Figure 4).

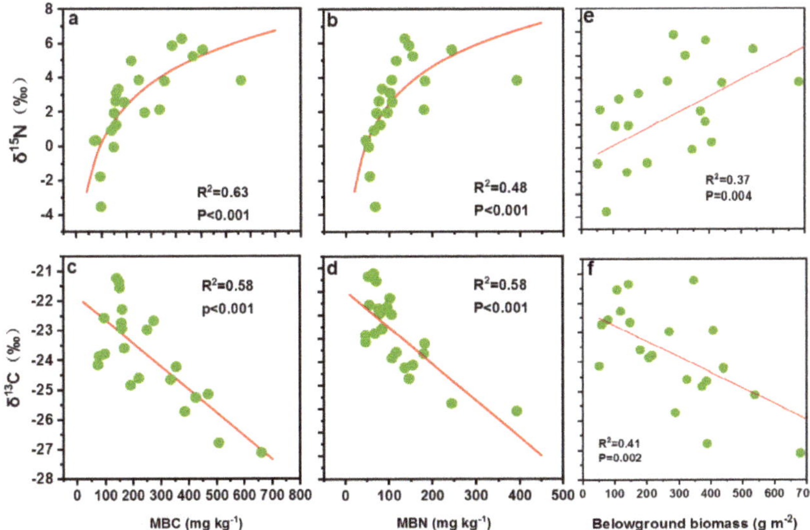

Figure 3. Relationships between soil δ^{15}N and δ^{13}C in the upper 5 cm with microbial biomass carbon (MBC, **a**, **c**) and microbial biomass nitrogen (MBN, **b**, **d**) as well as with plant belowground biomass (**e**, **f**) at 22 sites along the vegetation transect across Inner Mongolian temperate grasslands.

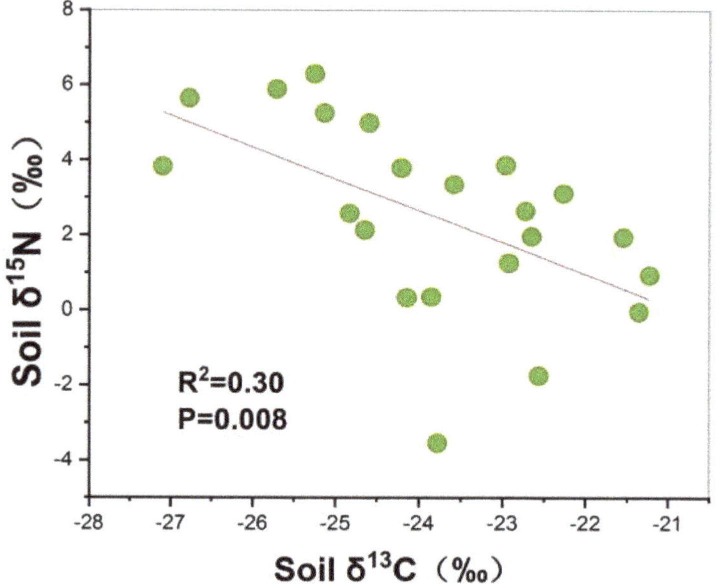

Figure 4. Relationships between soil δ^{15}N and δ^{13}C in the upper 5 cm at 22 sites along a vegetation transect across Inner Mongolian temperate grasslands.

The result of model averaging approach after model selection showed that MBC, total N, total P and SOC significantly positive related with soil δ^{15}N. Microbial C, total P, and MAT were three most important factors which significantly negatively affected soil δ^{13}C (Table 3).

Table 3. Relative importance and regression coefficients (given in parenthesis) of microbial biomass carbon (MBC), total nitrogen (N), mean annual precipitation (MAP), total phosphorous (P), soil organic carbon (SOC), and mean annual temperature (MAT) for determining soil ^{15}N and ^{13}C values. Values were derived through a model averaging approach. Factors with parameter values highlighted in bold significantly affected soil ^{15}N and ^{13}C ($P < 0.05$).

	MBC	Total N	MAP	Total P	SOC	MAT
Soil ^{15}N	**0.52** (0.012)	**0.21** (1.398)	0.20 (−0.007)	**0.16** (8.559)		**0.12** (0.119)
Soil ^{13}C		**0.29** (−0.006)	0.29 (−0.005)	**0.71** (−3.420)		**0.71** (−1.152)

4. Discussion

Numerous studies have explored the effects of various factors on soils δ^{13}C and δ^{15}N in terrestrial ecosystems on different scales [9,11,18,27,29,54]. However, it is still unclear how co-varying climatic, edaphic and biotic factors control soils δ^{13}C and δ^{15}N in temperate grasslands on a large scale. To answer the above question, a large-scale soil collection across temperate grasslands was carried out along a vegetation transect in Inner Mongolia. We found that biological and edaphic factors such as MBC and total N exert more effects on soil δ^{15}N whereas climatic and edaphic factors such as MAT and total P have more impacts on soil δ^{13}C.

Other studies have found that variations in soil δ^{15}N values largely depend on isotopic signatures of inputs and outputs, the input–output balance, N transformation and their specific isotope effects [29]. The factors affecting the above-mentioned processes can impact soil δ^{13}C and δ^{15}N signatures. In the current study, soil ^{15}N values across the transect had positive values, with the exception of two sites. Higher soil δ^{15}N values in drier ecosystems reflects a larger loss of mineral N through strongly

[15]N-discriminating processes, e.g., higher gaseous N losses caused by N-cycling microbes [34,55] and increased N mineralization and nitrification [56]. Numerous studies have suggested that climatic factors control δ^{15}N in soil [10,11,34,57]. In contrast to previous studies, the current study did not demonstrate a clear relationship between δ^{15}N and MAT or MAP [10,11,34]. These different findings could be ascribed to two reasons: (1) The length of our transect (i.e., the scale) was much short than those in both previous studies in the same region [10,34]. (2) As an indicator that integrates many processes affecting the N cycle, controls on soil δ^{15}N are very complicated, including various climatic, edaphic and biological factors [29,32]. Consistently with a previous study on the Chinese Loess Plateau [58], edaphic factors such as SOC, N_t and P_t strongly influenced soil δ^{15}N across our transect. Previous studies demonstrated that soil δ^{15}N increased or decreased with N_t contents [10,11], but we observed a significant positive logarithmic relationship between soil δ^{15}N and N_t along the transect. Among the various factors, MBC, i.e., a biotic factor, played a more important role in controlling the soil ^{15}N signature than climatic and edaphic factors (Table 2). This reflects that microbial processes are responsible for soil ^{15}N dynamics across the investigated temperate grassland, supporting our first hypothesis.

Previous studies have shown that soil δ^{13}C signature corresponds similarly to biotic factors, such as plant residue input from litterfall and rhizodeposition, including root mortality and root exudation [59,60]. Over time, dynamics of soil δ^{13}C are therefore largely controlled by C inputs from vegetation and subsequent microbial decomposition [13,18,61]. However, we found that climate and edaphic properties exerted greater control on soil ^{13}C in the investigated temperate grasslands (Table 2), which is consistent with previous studies demonstrating the importance of climate on soil δ^{13}C [13,62]. Additionally, a previous study also showed that the spatial variation of soil δ^{13}C was related to soil texture in a subtropical lowland woodland [25]. Considering that δ^{13}C can be regarded as an indicator of SOC dynamics, our results suggest that SOC dynamics in temperate grasslands are largely controlled by climatic and edaphic factors since MAP and P_t most dominantly affected soil ^{13}C values. Therefore, our results confirm our second hypothesis that climatic and edaphic factors have a higher effect on soil δ^{13}C than biological factors. This could be because water is a critical factor limiting growth of plants and microorganisms in these arid and semi-arid temperate grasslands [63,64]. Additionally, P is a key nutrient in temperate grasslands and, together with N, co-limits plant net primary production and microbial activities [65,66]. Therefore, both precipitation and P_t affect soil ^{13}C values by altering C input and microbial decomposition. These findings indicate that climatic and edaphic factors should be taken into account in order to better understand SOC dynamics, especially focusing on their roles in the microbial decomposition of plant residues and SOC.

5. Conclusions

In summary, a large-scale soil collection was carried out in temperate grasslands along a vegetation transect in Inner Mongolia to examine how climatic, edaphic and biological factors affect the soil δ^{13}C or δ^{15}N signature. Along the transect, soil δ^{15}N was not correlated with MAP and MAT, while soil δ^{13}C linearly decreased with MAP and MAT. Soil δ^{15}N logarithmically increased with concentrations of SOC, N_t and P_t, but soil δ^{13}C linearly decreased with concentrations of SOC, N_t and P_t. Similarly, soil δ^{15}N logarithmically increased with MBC and MBN, while soil δ^{13}C linearly decreased with MBC and MBN. Soil δ^{15}N linearly increased with plant belowground biomass, but soil δ^{13}C decreased with plant belowground biomass. Multiple linear regressions showed that MBC especially, but also N_t to a lesser extent, affect soil δ^{15}N, while MAT and P_t have more impact on soil δ^{13}C. Thus, biotic factors controlled soil ^{15}N signature most dominantly while climatic and edaphic factors exerted greater control on soil δ^{13}C signature. These results indicate that soil C and N cycles are to some extent decoupled in these temperate grasslands. Further investigations should focus on those ecological processes leading to decoupling of C and N cycles in temperate grassland soils for a better understanding of SOC dynamics [12].

Author Contributions: X.C. and X.X. conceived and designed the experiments; X.Z. and F.W. performed the experiments; X.Z., M.L. and L.Z. analyzed the data; X.C. contributed reagents/materials/analysis tools; X.Z. wrote the paper; X.C., X.X., R.C. and I.G. contributed to interpret the results and improve the manuscript. All authors have read and agreed to the published version of the manuscript.

Funding: This research was funded by the National Key Research and Development Program of China (2016YFC0501802; 2017YFA0604802) and by the National Natural Science Foundation of China (41877089).

Acknowledgments: This study was supported by the National Key Research and Development Program of China (Grant No. 2016YFC0501802; 2017YFA0604802).

Conflicts of Interest: The authors declare no conflict of interest.

References

1. Miltner, A.; Bombach, P.; Schmidt-Bruecken, B.; Kaestner, M. SOM genesis: Microbial biomass as a significant source. *Biogeochemistry* **2012**, *111*, 41–55. [CrossRef]
2. Schnitzer, M. A lifetime perspective on the chemistry of soil organic matter. *Adv. Agron.* **1999**, *68*, 1–58. [CrossRef]
3. Batjes, N.H. Total carbon and nitrogen in the soils of the world. *Eur. J. Soil Sci.* **1996**, *47*, 151–163. [CrossRef]
4. Post, W.M.; Pastor, J.; Zinke, P.J.; Stangenberger, A.G. Global patterns of soil nitrogen storage. *Nature* **1985**, *317*, 613–616. [CrossRef]
5. Janzen, H.H.; Campbell, C.A.; Ellert, B.H.; Bremer, E. Soil organic matter dynamics and their relationship to soil quality. *Dev. Soil Sci.* **1997**, *25*, 277–291. [CrossRef]
6. Schmidt, M.W.I.; Torn, M.S.; Abiven, S.; Dittmar, T.; Guggenberger, G.; Janssens, I.A.; Kleber, M.; Kögel-Knabner, I.; Lehmann, J.; Manning, D.A.C.; et al. Persistence of soil organic matter as an ecosystem property. *Nature* **2011**, *478*, 49–56. [CrossRef]
7. Goh, K.M. Carbon sequestration and stabilization in soils: Implications for soil productivity and climate change. *Soil Sci. Plant Nutr.* **2004**, *50*, 467–476. [CrossRef]
8. Lal, R. Managing soils and ecosystems for mitigating anthropogenic carbon emissions and advancing global food Security. *Bioscience* **2010**, *60*, 708–721. [CrossRef]
9. Amundson, R.; Austin, A.T.; Schuur, E.A.G.; Yoo, K.; Matzek, V.; Kendall, C.; Uebersax, A.; Brenner, D.L.; Baisden, W.T. Global patterns of the isotopic composition of soil and plant nitrogen. *Glob. Biogeochem. Cycles* **2003**, *17*, 1031. [CrossRef]
10. Cheng, W.X.; Chen, Q.S.; Xu, Y.Q.; Han, X.G.; Li, L.H. Climate and ecosystem ^{15}N natural abundance along a transect of Inner Mongolian grasslands: Contrasting regional patterns and global patterns. *Glob. Biogeochem. Cycles* **2009**, *23*, GB2005. [CrossRef]
11. Craine, J.M.; Elmore, A.J.; Wang, L.; Augusto, L.; Baisden, W.T.; Brookshire, E.N.J.; Cramer, M.D.; Hasselquist, N.J.; Hobbie, E.A.; Kahmen, A.; et al. Convergence of soil nitrogen isotopes across global climate gradients. *Sci. Rep.* **2015**, *5*, 8280. [CrossRef]
12. Ehleringer, J.R.; Buchmann, N.; Flanagan, L.B. Carbon isotope ratios in belowground carbon cycle processes. *Ecol. Appl.* **2000**, *10*, 412–422. [CrossRef]
13. Garten, C.T.; Cooper, L.W.; Post, W.M.; Hanson, P.J. Climate controls on forest soil C isotope ratios in the southern appalachian mountains. *Ecology* **2000**, *81*, 1108–1119. [CrossRef]
14. Wang, C.; Houlton, B.Z.; Liu, D.; Hou, J.; Cheng, W.; Bai, E. Stable isotopic constraints on global soil organic carbon turnover. *Biogeosciences* **2018**, *15*, 987–995. [CrossRef]
15. *Stable Isotopes in Ecology and Environmental Science*, 2nd ed.; Michener, R.H.; Lajtha, K. (Eds.) Blackwell Publishing Ltd.: Oxford, UK, 2007; ISBN 978-1-4051-2680-9.
16. Bernoux, M.; Cerri, C.C.; Neill, C.; de Moraes, J.F.L. The use of stable carbon isotopes for estimating soil organic matter turnover rates. *Geoderma* **1998**, *82*, 43–58. [CrossRef]
17. Conen, F.; Zimmermann, M.; Leifeld, J.; Seth, B.; Alewell, C. Relative stability of soil carbon revealed by shifts in δ^{15}N and C:N ratio. *Biogeosciences* **2008**, *5*, 123–128. [CrossRef]
18. Natelhoffer, K.J.; Fry, B. Controls on natural ^{15}N and ^{13}C abundances in forest soil organic matter. *Soil Sci. Soc. Am. J.* **1988**, *52*, 1633–1640. [CrossRef]
19. Bai, E.; Boutton, T.W.; Liu, F.; Wu, X.B.; Archer, S.R. Spatial patterns of soil δ^{13}C reveal grassland-to-woodland successional processes. *Org. Geochem.* **2012**, *42*, 1512–1518. [CrossRef]

20. Balesdent, J.; Mariotti, A. Measurement of soil organic matter turnover using ^{13}C natural abundance. In *Mass Spectrometry of Soil*; Boutton, T.W., Yamasaki, S.I., Eds.; Marcel Dekker, Inc.: New York, NY, USA, 1996; pp. 83–111. ISBN 0-8247-9699-3.

21. Kramer, M.G.; Sollins, P.; Sletten, R.S.; Swart, P.K. N isotope fractionation and measures of organic matter alteration during decomposition. *Ecology* **2003**, *84*, 2021–2025. [CrossRef]

22. Biedenbender, S.H.; McClaran, M.P.; Quade, J.; Weltz, M.A. Landscape patterns of vegetation change indicated by soil carbon isotope composition. *Geoderma* **2004**, *119*, 69–83. [CrossRef]

23. Mayor, J.R.; Wright, S.J.; Schuur, E.A.; Brooks, M.E.; Turner, B.L. Stable nitrogen isotope patterns of trees and soils altered by long-term nitrogen and phosphorus addition to a lowland tropical rainforest. *Biogeochemistry* **2014**, *119*, 293–306. [CrossRef]

24. Stevenson, B.A.; Parfitt, R.L.; Schipper, L.A.; Baisden, W.T.; Mudge, P. Relationship between soil δ^{15}N, C/N and N losses across land uses in New Zealand. *Agric. Ecosyst. Environ.* **2010**, *139*, 736–741. [CrossRef]

25. Bai, E.; Boutton, T.W.; Liu, F.; Ben Wu, X.; Hallmark, C.T.; Archer, S.R. Spatial variation of soil δ^{13}C and its relation to carbon input and soil texture in a subtropical lowland woodland. *Soil Biol. Biochem.* **2012**, *44*, 102–112. [CrossRef]

26. Wynn, J.G.; Bird, M.I. Environmental controls on the stable carbon isotopic composition of soil organic carbon: Implications for modelling the distribution of C_3 and C_4 plants, Australia. *Tellus* **2008**, *60*, 604–621. [CrossRef]

27. Craine, J.M.; Brookshire, E.N.J.; Cramer, M.D.; Hasselquist, N.J.; Koba, K.; Marin-Spiotta, E.; Wang, L. Ecological interpretations of nitrogen isotope ratios of terrestrial plants and soils. *Plant Soil* **2015**, *396*, 1–26. [CrossRef]

28. Dawson, T.E.; Mambelli, S.; Plamboeck, A.H.; Templer, P.H.; Tu, K.P. Stable isotopes in plant ecology. *Annu. Rev. Ecol. Syst.* **2002**, *33*, 507–559. [CrossRef]

29. Högberg, P. ^{15}N natural abundance in soil-plant systems. *New Phytol.* **1997**, *137*, 179–203. [CrossRef]

30. Makarov, M.I. The nitrogen isotopic composition in soils and plants: Its use in environmental studies (a review). *Eurasian Soil Sci.* **2009**, *42*, 1335–1347. [CrossRef]

31. Pardo, L.H.; Nadelhoffer, K.J. Using nitrogen isotope ratios to assess terrestrial ecosystems at regional and global scales. In *Isoscapes: Understanding Movement, Pattern, and Process on Earth through Isotope Mapping*; West, J.B., Bowen, G.J., Dawson, T.E., Tu, K.P., Eds.; Springer: New York, NY, USA, 2010; pp. 221–249. ISBN 978-90-481-3353-6.

32. Robinson, D. δ^{15}N as an integrator of the nitrogen cycle. *Trends Ecol. Evol.* **2001**, *16*, 153–162. [CrossRef]

33. Powers, J.S.; Veldkamp, E. Regional variation in soil carbon and δ^{13}C in forests and pastures of northeastern Costa Rica. *Biogeochemistry* **2005**, *72*, 315–336. [CrossRef]

34. Wang, C.; Wang, X.; Liu, D.; Wu, H.; Lu, X.; Fang, Y.; Cheng, W.; Luo, W.; Jiang, P.; Shi, J.; et al. Aridity threshold in controlling ecosystem nitrogen cycling in arid and semi-arid grasslands. *Nat. Commun.* **2014**, *5*, 4799. [CrossRef]

35. Handley, L.L.; Raven, J.A. The use of natural abundance of nitrogen isotopes in plant physiology and ecology. *Plant Cell Environ.* **1992**, *15*, 965–985. [CrossRef]

36. Aranibar, J.N.; Otter, L.; Macko, S.A.; Feral, C.J.W.; Epstein, H.E.; Dowty, P.R.; Eckardt, F.; Shugart, H.H.; Swap, R.J. Nitrogen cycling in the soil-plant system along a precipitation gradient in the Kalahari sands. *Glob. Chang. Biol.* **2004**, *10*, 359–373. [CrossRef]

37. Austin, A.T.; Vitousek, P.M. Nutrient dynamics on a precipitation gradient in Hawai'i. *Oecologia* **1998**, *113*, 519–529. [CrossRef]

38. Handley, L.L.; Austin, A.T.; Robinson, D.; Scrimgeour, C.M.; Raven, J.A.; Heaton, T.H.E.; Schmidt, S.; Stewart, G.R. The ^{15}N natural abundance (δ^{15}N) of ecosystem samples reflects measures of water availability. *Aust. J. Plant Physiol.* **1999**, *26*, 185–199. [CrossRef]

39. Qiao, N.; Xu, X.; Cao, G.; Ouyang, H.; Kuzyakov, Y. Land use change decreases soil carbon stocks in Tibetan grasslands. *Plant Soil* **2015**, *395*, 231–241. [CrossRef]

40. Wang, L.; D'Odorico, P.; Ries, L.; Macko, S.A. Patterns and implications of plant-soil δ^{13}C and δ^{15}N values in African savanna ecosystems. *Quat. Res.* **2010**, *73*, 77–83. [CrossRef]

41. Conant, R.T.; Cerri, C.E.P.; Osborne, B.B.; Paustian, K. Grassland management impacts on soil carbon stocks: A new synthesis. *Ecol. Appl.* **2017**, *27*, 662–668. [CrossRef]

42. Ramankutty, N.; Evan, A.T.; Monfreda, C.; Foley, J.A. Farming the planet: 1. Geographic distribution of global agricultural lands in the year 2000. *Glob. Biogeochem. Cycles* **2008**, *22*, GB1003. [CrossRef]

43. Foley, J.A.; Ramankutty, N.; Brauman, K.A.; Cassidy, E.S.; Gerber, J.S.; Johnston, M.; Mueller, N.D.; O'Connell, C.; Ray, D.K.; West, P.C.; et al. Solutions for a cultivated planet. *Nature* **2011**, *478*, 337–342. [CrossRef]

44. Climate Change 2014: Synthesis Report. *Contribution of Working Groups I, II and III to the Fifth Assessment Report of the Intergovernmental Panel on Climate Change*; Core Writing Team, Pachauri, R.K., Meyer, L.A., Eds.; IPCC: Geneva, Switzerland, 2014; ISBN 978-92-9169-143-2.

45. Bredenkamp, G.J.; Spada, F.; Kazmierczak, E. On the origin of northern and southern hemisphere grasslands. *Plant Ecol.* **2002**, *163*, 209–229. [CrossRef]

46. Farquhar, G.D.; Richards, R.A. Isotopic composition of plant carbon correlates with water use efficiency of wheat genotypes. *Aust. J. Plant Physiol.* **1984**, *11*, 539–552. [CrossRef]

47. Soil Survey Staff. *Keys to Soil Taxonomy*, 9th ed.; USDA-Natural Resources Conservation Service: Washington, DC, USA, 2003.

48. Grimshaw, H.M. The determination of total phosphorus in soils by acid digestion. In *Chemical Analysis in Environmental Research*; Rowland, A.P., Ed.; NERC/ITE: Abbots Ripton, UK, 1987; pp. 92–95. ISBN 0-904282-98-8.

49. Parkinson, J.A.; Allen, S.E. A wet oxidation procedure suitable for the determination of nitrogen and mineral nutrients in biological material. *Commun. Soil Sci. Plant Anal.* **1975**, *6*, 1–11. [CrossRef]

50. Brookes, P.C.; Landman, A.; Pruden, G.; Jenkinson, D.S. Chloroform fumigation and the release of soil nitrogen: A rapid direct extraction method to measure microbial biomass nitrogen in soil. *Soil Biol. Biochem.* **1985**, *17*, 837–842. [CrossRef]

51. Vance, E.D.; Brookes, P.C.; Jenkinson, D.S. An extraction method for measuring soil microbial biomass C. *Soil Biol. Biochem.* **1987**, *19*, 703–707. [CrossRef]

52. Wu, J.S.; Joergensen, R.G.; Pommerening, B.; Chaussod, R.; Brookes, P.C. Measurement of soil microbial biomass C by fumigation extraction—An automated procedure. *Soil Biol. Biochem.* **1990**, *22*, 1167–1169. [CrossRef]

53. Burnham, K.P.; Anderson, D.R.; Huyvaert, K.P. AIC model selection and multimodel inference in behavioral ecology: Some background, observations, and comparisons. *Behav. Ecol. Sociobiol.* **2011**, *65*, 23–35. [CrossRef]

54. Cheng, S.L.; Fang, H.J.; Yu, G.R.; Zhu, T.H.; Zheng, J.J. Foliar and soil ^{15}N natural abundances provide field evidence on nitrogen dynamics in temperate and boreal forest ecosystems. *Plant Soil* **2010**, *337*, 285–297. [CrossRef]

55. Xu, Y.; He, J.; Cheng, W.; Xing, X.; Li, L. Natural ^{15}N abundance in soils and plants in relation to N cycling in a rangeland in Inner Mongolia. *J. Plant Ecol.* **2010**, *3*, 201–207. [CrossRef]

56. Kahmen, A.; Wanek, W.; Buchmann, N. Foliar δ^{15}N values characterize soil N cycling and reflect nitrate or ammonium preference of plants along a temperate grassland gradient. *Oecologia* **2008**, *156*, 861–870. [CrossRef]

57. Weintraub, S.R.; Cole, R.J.; Schmitt, C.G.; All, J.D. Climatic controls on the isotopic composition and availability of soil nitrogen across mountainous tropical forest. *Ecosphere* **2016**, *7*, e01412. [CrossRef]

58. Shan, Y.; Huang, M.; Suo, L.; Zhao, X.; Wu, L. Composition and variation of soil δ^{15}N stable isotope in natural ecosystems. *Catena* **2019**, *183*, 104236. [CrossRef]

59. Nissenbaum, A.; Schallinger, K.M. The distribution of stable carbon isotope (^{13}C/^{12}C) in fractions of soil organic matter. *Geoderma* **1974**, *11*, 137–145. [CrossRef]

60. Vankessel, C.; Farrell, R.E.; Pennock, D.J. ^{13}C and ^{15}N natural abundance in crop residues and soil organic matter. *Soil Sci. Soc. Am. J.* **1994**, *58*, 382–389. [CrossRef]

61. Santruckova, H.; Bird, M.I.; Lloyd, J. Microbial processes and carbon-isotope fractionation in tropical and temperate grassland soils. *Funct. Ecol.* **2000**, *14*, 108–114. [CrossRef]

62. Du, B.; Liu, C.; Kang, H.; Zhu, P.; Yin, S.; Shen, G.; Hou, J.; Ilvesniemi, H. Climatic control on plant and soil delta C-13 along an altitudinal transect of Lushan mountain in subtropical China: Characteristics and interpretation of soil carbon dynamics. *PLoS ONE* **2014**, *9*, e86440. [CrossRef]

63. Guo, Q.; Hu, Z.; Li, S.; Li, X.; Sun, X.; Yu, G. Spatial variations in aboveground net primary productivity along a climate gradient in Eurasian temperate grassland: Effects of mean annual precipitation and its seasonal distribution. *Glob. Chang. Biol.* **2012**, *18*, 3624–3631. [CrossRef]

64. Jiao, C.; Yu, G.; He, N.; Ma, A.; Ge, J.; Hu, Z. Spatial pattern of grassland aboveground biomass and its environmental controls in the Eurasian steppe. *J. Geogr. Sci.* **2017**, *27*, 3–22. [CrossRef]
65. Bai, X.; Cheng, J.; Zheng, S.; Zhan, S.; Bai, Y. Ecophysiological responses of *Leymus chinensis* to nitrogen and phosphorus additions in a typical steppe. *Chin. J. Plant Ecol.* **2014**, *38*, 103–115. [CrossRef]
66. Dong, C.; Wang, W.; Liu, H.; Xu, X.; Zeng, H. Temperate grassland shifted from nitrogen to phosphorus limitation induced by degradation and nitrogen deposition: Evidence from soil extracellular enzyme stoichiometry. *Ecol. Indic.* **2019**, *101*, 453–464. [CrossRef]

Article

Climate-Driven Holocene Migration of Forest-Steppe Ecotone in the Tien Mountains

Ying Cheng [1], Hongyan Liu [2,*], Hongya Wang [2], Qian Hao [3], Yue Han [2], Keqin Duan [1] and Zhibao Dong [1]

[1] School of Geography and Tourism, Shaanxi Normal University, Xi'an 710119, China; chengyingnature@163.com (Y.C.); kqduan@snnu.edu.cn (K.D.); zbdong@snnu.edu.cn (Z.D.)

[2] College of Urban and Environmental Sciences and MOE Laboratory for Earth Surface, Peking University, Beijing 100871, China; why@urban.pku.edu.cn (H.W.); hanyuegeo@foxmail.com (Y.H.)

[3] Institute of Surface-Earth System Science, School of Earth System Science, Tianjin University, Tianjin 300072, China; haoqian@tju.edu.cn

* Correspondence: lhy@urban.pku.edu.cn

Received: 7 September 2020; Accepted: 26 October 2020; Published: 28 October 2020

Abstract: Climate change poses a considerable threat to the forest-steppe ecotone in arid mountain areas. However, it remains unclear how the forest-steppe ecotone responds to climate change due to the limitation of the traditional pollen assemblages, which greatly limits the understanding of the history of the forest-steppe ecotone. Here, we examined the Tien Mountains, the largest mountain system in the world's arid regions, as a case study to explore the migration of the forest-steppe ecotone using the pollen taxa diversity, by combining modern vegetation surveys, surface pollen and two fossil pollen sequences—in the mid-elevation forest belt (Sayram Lake) and in the low-elevation desert belt (Aibi Lake). We found that the forest-steppe migration followed Holocene climate change. Specifically, the forest belt where *Picea schrenkiana* Fisch. & C.A.Mey. dominates has a very low pollen taxa diversity, characterized by high richness and low evenness, which plays a key role in mountainous diversity. By detecting the diversity change of the deposition sites, we found that in coping with the warm and wet middle Holocene, the forest belt expanded and widened as the observed diversities around the two lakes were very low, thus the forest-steppe ecotone moved downward accordingly. During the early and late Holocene, the forest belt and the forest-steppe ecotone moved upward under a warm and dry climate, and downward under a cold and wet climate, as there was a reduced forest belt effect on, or contribution to, the sites, and the observed diversities were high. Moisture loss may pose the greatest threat to the narrow forest-steppe ecotone, considering the climatic niche space and the limited living space for humidity-sensitive taxa. This study highlights that temperature and moisture co-influence the forest belt change, which further determines the position migration of the forest-steppe ecotone.

Keywords: pollen analysis; forest belt; forest-steppe ecotone; position migration; moisture change

---

1. Introduction

Arid and semi-arid areas are extremely sensitive to climate change and human activities, and the forest-steppe ecotone in these areas is among the most vulnerable ecosystems in the world [1–3]. Until now, climate change has posed a serious threat to the forest-steppe ecotone in arid and semi-arid areas by causing tree growth decline and tree mortality [2,4,5]. The latest forecasts suggest that the expansion of arid and semi-arid areas will potentially exacerbate regional warming and habitat fragmentation in the future [3,6]. Therefore, there is an urgent need to pay attention to the forest-steppe ecotone in such ecologically fragile areas.

Pollen taxa concentrates in mountains in arid regions, which are also key areas for ecosystem services. Mountain areas typically have rich and unique pollen taxa [7]. Generally, mountain vegetation belts in arid central Asia have various types, ranging from desert to grassland, forest and alpine meadows as the elevation rises [8]. Among them, mountain forest-steppes, typically situated between the forest and steppe belts, are located at the driest edge of forest distribution vertically [9,10]. As such, it is critical to understand the dynamics of the mountain forest-steppe ecotone under climate change in arid regions. Yet, how the change of pollen reflects the migration of the forest-steppe ecotone remains unclear, despite the series of traditional pollen assemblage studies carried out in some hotspots, including the Tibetan Plateau, the Altai Mountains, and the Taibai Mountains [11–13].

Pollen taxa diversity can serve as a comprehensive indicator reflecting the dynamics of the vegetation belts [4]. Whether past climate change increased or decreased pollen taxa diversity has not yet been fully determined, and changes in the pollen taxa diversity potentially include clues for vegetation belt change, especially for the migration of the mountain forest-steppe ecotone. Thus, the Holocene is an ideal period to explore the migration of the forest-steppe ecotone, wherein which climate change was mainly caused by natural variability.

Moreover, it is important to focus on the response modes of mountain vegetation belts to Holocene climate change and to improve the corresponding mechanisms, which are the key to exploring the history of forest-steppe migration. The position of the mountainous vertical vegetation belt is very sensitive to climate change [11,12], because the vertical belt is the epitome of the horizontal belt (about 1/1000). The vertical belt in arid areas is unique, as the upper and lower limits of the forest are affected by temperature and moisture, respectively. Mountain vegetation responds to temperature change by position migration, while it responds to moisture change through changes in species composition [4,5]. These two modes are mostly adopted, even though they might not fully explain the mechanism between mountain vegetation belts and changes in temperature and moisture. Studies have indicated that the width of the mountain vegetation belt and climate are not completely in equilibrium [14,15]. As the climate mode can be warm and dry, warm and wet, cold and dry, and cold and wet, both the upper and the lower limits of the forest belt, which are in different climatic conditions, may respond differently [12], which ultimately leads to potential changes in the belt width and position. As such, variations in the width of vegetation belts may also be one of the most typical responses to particular climate change, and may further lead to the change in the forest-steppe ecotone. Therefore, it is of importance to see whether these kinds of response patterns exist in the vegetation belts and can be detected through the change of pollen taxa diversity in arid mountain areas.

The Tien Mountains, the largest mountain system in the world's arid regions, is located in the hinterland of Eurasia, affected by the westerlies [16–18], where the vertical differentiation of vegetation belts is apparent and the typical mountain forest-steppe exists [19]. Due to the complex topographical conditions, limited range size and arid environments, the forest-steppe ecotone in the Tien Mountains appears to be most threatened by climate deterioration. However, few studies have specifically focused on the migration of the forest-steppe ecotone in response to the effect of climate change in the Tien Mountains, despite many studies exploring its climate and vegetation changes at multiple time scales [20–23]. Therefore, there is an urgent need to comprehensively study the response pattern of the forest-steppe ecotone in the Tien Mountains to Holocene climate change based on evidence from both surface pollen and fossil pollen, as pollen can provide a long history of pollen taxa diversity with good temporal continuity [4].

Based on the above description, we aim to explore the following scientific questions: (1) How did the mountainous pollen taxa diversity evolve during the Holocene? (2) How did the pollen taxa diversity reflect the changes in vegetation belts in the Tien Mountains during the Holocene? (3) How did the forest-steppe ecotone migrate in response to the Holocene temperature and moisture change in the Tien Mountains?

2. Materials and Methods

2.1. Study Area

The Tien Mountains, the largest mountain system in the world's arid regions, are located in the hinterland of Eurasia, influenced by westerlies (Figure 1a,b) [16]. The annual mean precipitation on the northern slopes changes from 175 mm in the desert zone to 465 mm in the mid-altitude zone [24].

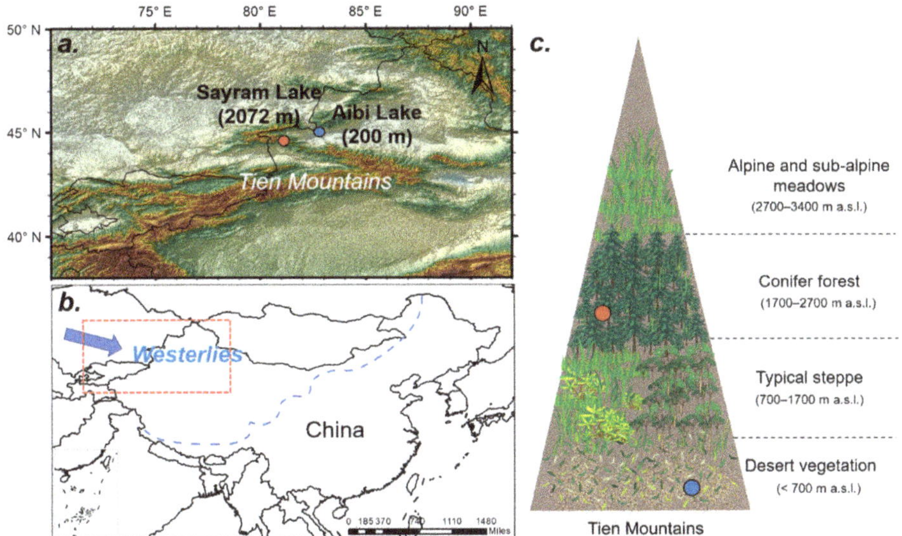

Figure 1. The location of the Tien Mountains in the hinterland of Eurasia, influenced by westerly winds. (**a**) Digital elevation model image of the study area. The red solid dot represents the location of Sayram Lake, while the blue one represents the location of Aibi Lake. (**b**) The red dotted area represents the range of subgraph. The blue dotted line represents the modern limit of the East Asian summer monsoon, which also illustrates the horizontal limit of the westerlies effect. (**c**) Vegetation belts on the northern slope of the Tien Mountains.

The Tien Mountains have the most typical and complete vertical natural belt spectrum for temperate arid areas around the world, reflecting the distribution characteristics and changes in the mountain pollen taxa diversity and ecological processes. The northern slope of the Tien Mountains can be divided into the following vegetation belts [14] (Figure 1c): sparse and low cushion vegetation is distributed in high mountain belts above 3400 m a.s.l.; alpine and subalpine meadows dominate from 2700 to 3400 m a.s.l.; conifer forest is distributed on the middle elevations from 1700 to 2700 m a.s.l.; typical steppe dominates from 700 to 1700 m a.s.l., represented by steppe on the sunny slope and patchy forest on the shady slope; desert vegetation is distributed below 700 m a.s.l. Notably, the forest-steppe ecotone, also known as the transition zone between the forest belt and the steppe belt, is currently situated around 1700 m a.s.l. on the northern slope of the Tien Mountains [14].

2.2. Modern Vegetation Survey

To investigate the changes in the richness of modern plant species along the elevational gradient, typical plots representing the characteristic vegetation composition in each vegetation belt were selected, based on the natural geographical pattern of the Tien Mountains [14]. In the mountainous area, with an elevation from 700 to 1700 m a.s.l., plots were sampled every 100 m along the elevational gradient. Due to a flat terrain in the oasis and the downstream desert area, plots were sampled every

50 km in the horizontal direction. Finally, 40 valid plots were surveyed. During the sample survey, the forest plot area was 10 m × 10 m, and the grassland or meadow community area was 2 m × 2 m. Then the detailed information about vegetation habitat conditions, species composition, coverage and height were recorded.

2.3. Surface Pollen

Surface soil samples were collected every 100 m in the range of 200 to 3500 m a.s.l. on the northern slope of the Tien Mountains, and 34 samples were obtained to investigate the changes in surface pollen composition and evenness in each vegetation belt [14]. Pollen was extracted by treating with HCl (10%) and NaOH (10%), then sieved through 180 μm and 10 μm sieves, and floated with heavy liquid (KI + HI + Zn) using standard techniques [25]. Each sample was counted to contain at least 250 pollen grains under an optical microscope at 400× magnification.

2.4. Fossil Pollen

We selected two typical fossil pollen sites with high resolution on the northern slope of the Tien Mountains to explore the evolution of pollen taxa diversity. We digitized all pollen taxa using GetData Graph Digitizer v2.25 software [26] and then used them as analytical paleo data. Considering the location representativeness, age, temporal resolution and pollen identification of fossil pollen cores, the following two sites can meet our analysis needs.

The first sediment core, named SLMH-2009 (44°35′ N, 81°09′ E), was from Sayram Lake, with an elevation of 2072 m (Figure 1a,b) [22], and was located in the middle-low part of the modern conifer forest belt. This core was dated back to the Late Glacial to early Holocene transition, with 11 AMS^{14}C dates, and had a temporal resolution of less than 100 years for each pollen sample. The authors analyzed 150 fossil pollen samples in total, with a minimum of 400 pollen grains identified for each sample.

The second sediment core, named A-01, was from Aibi Lake (44°54′–45°08′ N, 82°35′–83°10′ E) (Figure 1a,b) [23], located in the desert vegetation belt at 200 m a.s.l. This core was dated back to the Late Glacial to early Holocene transition with eight AMS^{14}C dates and had a temporal resolution of less than 150 years for each pollen sample. The authors analyzed 195 fossil pollen samples in total, with a minimum of 350 pollen grains identified for each sample.

2.5. Pollen Taxa Diversity Index

We chose the Shannon-Wiener index and Simpson index to calculate the pollen taxa diversity, as these two indices include the heterogeneity of the measured plant community. They consider both the richness and evenness of the species in the community. Specifically, the Shannon-Wiener diversity index was used to estimate the level of pollen taxa diversity in the modern ecosystem and in the Holocene, which is a comprehensive index reflecting richness and evenness. Its formula is as follows [27].

$$H = -\sum_{i=1}^{r} Pi \ln Pi \tag{1}$$

$Pi = ni/N$, indicating the relative richness of species.

ni represents the number of individuals of each pollen taxon.

N represents the total number of individuals of all pollen taxa in the community.

It should be noted that communities with low richness and high evenness, and communities with high richness and low evenness, have low diversity indices.

The Simpson diversity index was also used to present the pollen taxa diversity. Unlike the Shannon-Wiener index, the Simpson index measures the probability that two random entities from the pollen group represent different pollen taxa, which can be quantified by the following equation [28].

$$D = 1 - \sum_{i=1}^{N} P_i^2 \qquad (2)$$

$Pi = ni/N$, indicating the relative richness of species.

ni represents the number of individuals of each pollen taxon.

N represents the total number of individuals of all pollen taxa in the community.

2.6. Climate Data

Holocene temperature change was acquired from the integrated results of paleoclimate records for 30–90° N [29,30], and Holocene moisture change was derived from integrated results of paleoclimate records in arid central Asia [16].

3. Results

3.1. Modern Plant Species Richness Changes along the Elevational Gradient

We observed an increasing richness of modern plant species from low-elevation areas to mid-elevation mountainous areas (Figure 2). Regarding the oasis and the downstream desert area from 200 to 700 m a.s.l., plant species richness was limited, ranging from 4 to 17 (Figure 2). The range of 200 to 400 m a.s.l. belongs to the oasis area, with a relatively high number of species; therefore, the observed plant species greatly fluctuated (Figure 2). From 700 m a.s.l. to the middle elevation, plant species richness showed an increasing trend, with a peak occurring in the forest belt at 1700 m a.s.l. (Figure 2). According to our field survey, the wet-preferring species *Picea schrenkiana* Fisch. & C.A.Mey. was dominating the forest belt, and the richness of the forest belt was mainly caused by the species richness of the understory, which can reach nearly 40 species (Figure 2).

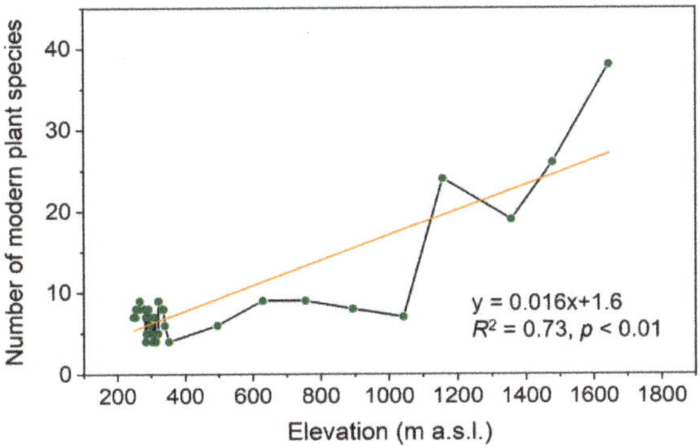

Figure 2. The richness of modern plant species along the elevational gradient. The orange line represents the linear regression line.

3.2. Surface Pollen Composition

The surface pollen taxa within the range of 250–3500 m on the northern slope of the Tien Mountains were analyzed according to their richness changes along the elevational gradient (Figure 3). *Salix*, *Ephedra*, Caryophyllaceae, Asteraceae, *Thalictrum*, Ranunculaceae, and Rosaceae peaked in the range of 3500–2700 m. *Picea*, Poaceae, and Ranunculaceae reached a peak in the range of 2700–1700 m.

Betula, Artemisia, Salix, Tamarix, Ephedra, Thalictrum, and Rosaceae peaked between 1700 and 700 m. *Ulmus,* Amaranthaceae, *Nitraria,* and *Tamarix* reached a peak below 700 m.

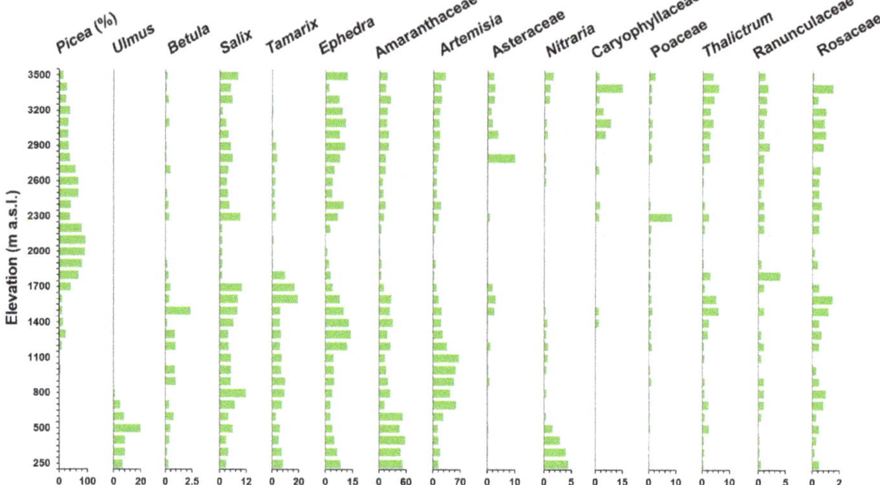

Figure 3. The richness of surface pollen taxa along the elevational gradient.

These taxa were further divided into the following four vegetation belts (Table 1). Alpine and sub-alpine meadows mainly included *Salix, Ephedra,* Caryophyllaceae, Asteraceae, *Thalictrum,* Ranunculaceae, and Rosaceae. These taxa can endure cold climates in the alpine zone. Conifer forest mainly included *Picea,* Poaceae and Ranunculaceae, where *Picea schrenkiana* was the dominant species favored by the wet climatic conditions. The typical steppe mainly contained *Betula, Artemisia, Salix, Tamarix, Ephedra, Thalictrum,* and Rosaceae. *Betula fruticosa* Pallas was located in the desert belt in front of the mountain, and is categorized as azonal vegetation. Desert vegetation contained *Ulmus,* Amaranthaceae, *Nitraria* and *Tamarix. Ulmus pumila* L. was distributed in the low-elevation river valleys, also belonging to the azonal vegetation classification. Amaranthaceae, *Nitraria,* and *Tamarix* are typically drought-tolerant species. Based on our division criteria, the richness of fossil pollen in the studied sedimentary site probably originated from different vegetation belts, which is soon discussed.

Table 1. Lists of main surface pollen taxa assigned to different vegetation belts along the northern slope of the Tien Mountains.

Alpine and Sub-Alpine Meadows 3500–2700 m a.s.l.	Conifer Forest 2700–1700 m a.s.l.	Typical Steppe 1700–700 m a.s.l.	Desert Vegetation Below 700 m a.s.l.
Salix		*Betula*	
Ephedra		*Artemisia*	
Caryophyllaceae	*Picea*	*Salix*	*Ulmus*
Asteraceae	Poaceae	*Tamarix*	Amaranthaceae
Thalictrum	Ranunculaceae	*Ephedra*	*Nitraria*
Ranunculaceae		*Thalictrum*	*Tamarix*
Rosaceae		Rosaceae	

Moreover, in the modern ecosystem, the Shannon-Wiener index and the Simpson index consistently indicated the lowest pollen taxa diversity in the forest belt, while the highest pollen taxa diversity was in the alpine and sub-alpine meadows belt, followed by the typical steppe belt, and the desert vegetation belt (Table 2). According to the result of the Shannon-Wiener index, in the upper (2800–2600 m) and lower (1800–1600 m) boundaries and the center of the forest belt (2400–2100 m), the pollen taxa diversities were 1.49, 1.65, and 1.21, respectively (Table 2). According to the result of the Simpson index, in the upper (2800–2600 m) and lower (1800–1600 m) boundaries and the center of the forest belt (2400–2100 m), the pollen taxa diversities were 0.63, 0.70, and 0.50, respectively (Table 2). Based on the results of the surface pollen analyses, we inferred that the pollen taxa diversity of the forest belt was generally low in the Tien Mountains, which was related to the very low pollen evenness due to the absolute dominance of *Picea* pollen in the forest belt (varying from 39% to 93%, with a mean of 67%) (Figures 2 and 3). Therefore, we inferred that the forest belt with very low pollen taxa diversity might affect the changes in pollen taxa diversity observed by the deposition sites during the Holocene, even though the forest belt had a larger plant species number than the belts of the typical steppe and desert vegetation (Figure 3).

Table 2. Vegetation belt and the pollen taxa diversity in the modern ecosystem.

Vegetation Belt	Altitude (m a.s.l.)	Pollen Taxa Diversity Indicated by the Shannon-Wiener Index	Pollen Taxa Diversity Indicated by the Simpson Index
Alpine and sub-alpine meadows	3500–2700	1.84	0.78
Transition zone (forest upper boundary)	2800–2600	1.49	0.63
Conifer forest	2700–1700	1.16	0.49
The center of the forest belt	2400–2100	1.21	0.50
Transition zone (forest lower boundary)	1800–1600	1.65	0.70
Typical steppe	1700–700	1.68	0.72
Desert vegetation	Below 700	1.52	0.67

3.3. Pollen Taxa Diversity Change around Sayram Lake during the Holocene

During the Holocene, belts of alpine and sub-alpine meadows, conifer forest and desert vegetation contributed to the pollen taxa diversity around Sayram Lake, while the belt of the typical steppe did not contribute, according to the regression relationships between pollen taxa diversity indicated by the Shannon-Wiener index and pollen richness from different vegetation belts (Figure 4). The linear regression showed that the conifer forest belt contributed the most to the pollen taxa diversity around Sayram Lake through the Holocene ($R^2 = 0.40$, $p < 0.01$), while contributions from the belts of alpine and sub-alpine meadows and desert vegetation were relatively small ($R^2 = 0.25$, $p < 0.01$ and $R^2 = 0.37$, $p < 0.01$, respectively) (Figure 4). This further suggests that pollen taxa diversity around Sayram Lake can reflect the forest belt change during the Holocene.

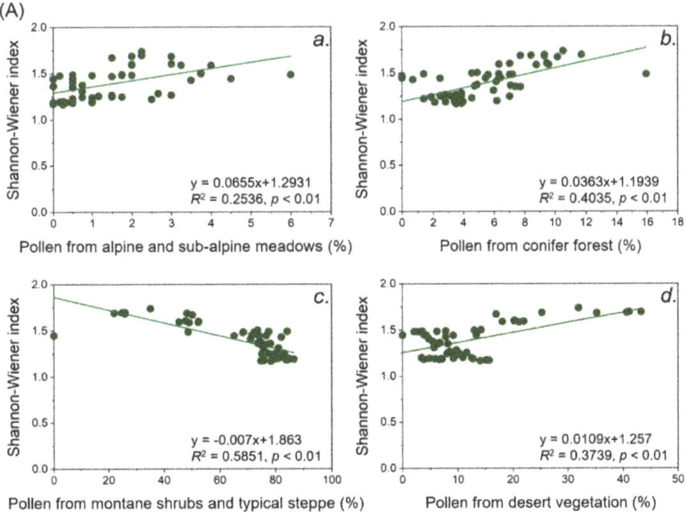

Figure 4. *Cont.*

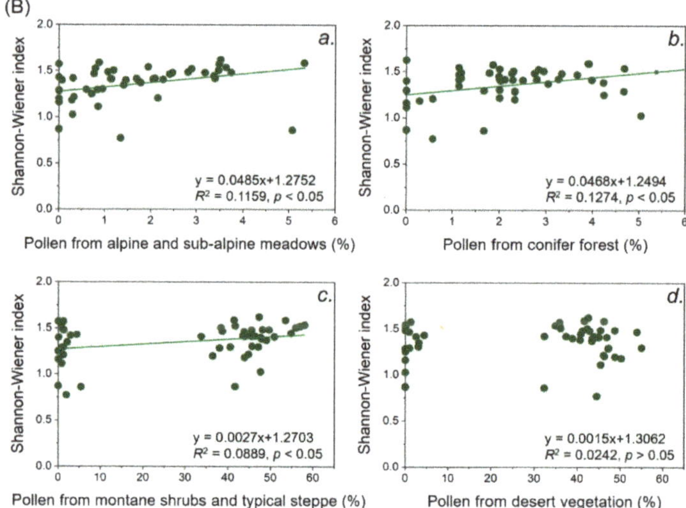

Figure 4. Linear relationships between the observed pollen percentage from different vegetation belts and the Shannon-Wiener index. Notably, the horizontal axis represents the sum of the percentage of pollen taxa derived from different vegetation belts (from a to d) observed in the sedimentary sites (specific taxa contained in each vegetation belt are listed in Table 1). Panel (**A**) represents Sayram Lake, and panel (**B**) represents Aibi Lake.

During the warm and dry periods of the early Holocene, from 12,000 to 8000 years before present (yr BP) (Figure 5a,b), the conifer forest belt moved up, and the belts of alpine and sub-alpine meadows and typical steppe moved up as well (Figure 6a), because the observed pollen taxa diversity indicated by the Shannon-Wiener index around Sayram Lake was very high, reaching a mean of 1.56 (Figure 5c), which was close to the diversity index (1.65) of the lower boundary of the modern forest belt (Table 2). Moreover, the observed pollen taxa diversity indicated by the Simpson index around Sayram Lake was also very high, reaching a mean of 0.69 (Figure 5d), which was close to the diversity index (0.70) of the lower boundary of the modern forest belt (Table 2). Therefore, we infer that Sayram Lake was probably at the lower boundary of the conifer forest during the early Holocene, thus the forest-steppe ecotone moved upward accordingly, and was close to Sayram Lake.

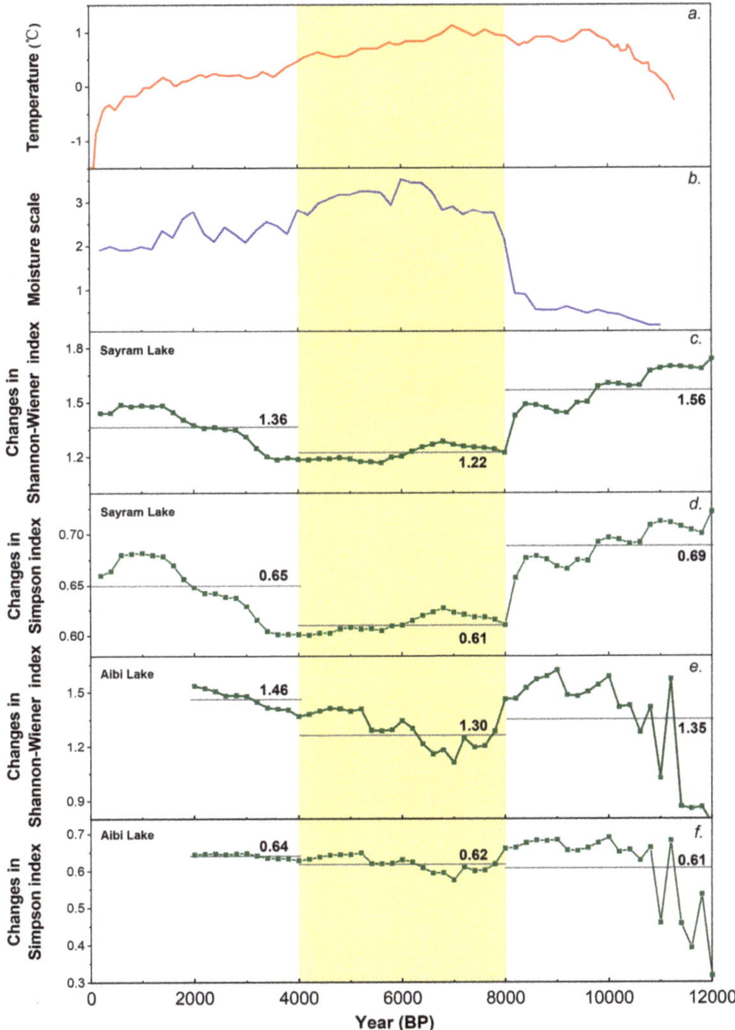

Figure 5. Changes in pollen taxa diversity in the Tien Mountains and the Holocene climate. (**a**) Holocene temperature change for 30–90° N [29,30]. (**b**) Holocene moisture changes in arid central Asia [16]. Gray lines represent mean levels of pollen taxa diversity index during different periods around Sayram Lake and Aibi Lake (**c**–**f**). The yellow area represents the middle Holocene with low pollen taxa diversity.

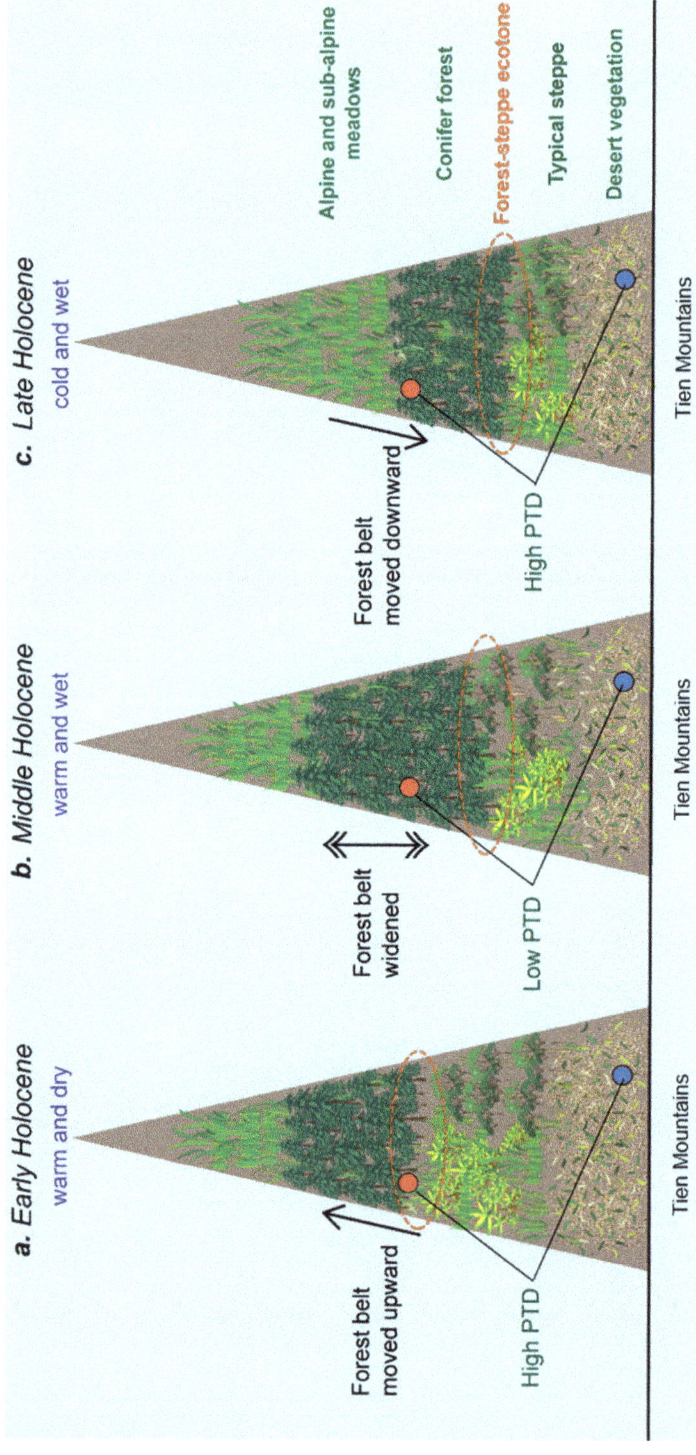

Figure 6. Schematic diagram presenting changes in pollen taxa diversity (PTD) and the migration of the forest-steppe ecotone under the Holocene's changing climate (**a–c**). In the belt of the typical steppe, the steppe is mainly distributed on the sunny slope, while the forest is distributed on the shady slope. The red dot represents the location of Sayram Lake, while the blue dot indicates the location of Aibi Lake. The orange ellipse indicates the possible range of the forest-steppe ecotone.

During the middle Holocene, from 8000 to 4000 yr BP (Figure 5a,b), when the climate was warm and wet compared to the last stage, the conifer forest belt expanded and dominated, while the belts of alpine and sub-alpine meadows and the typical steppe were conversely compressed (Figure 6b), as the observed pollen taxa diversity indicated by the Shannon-Wiener index around Sayram Lake was very low, reaching a mean of 1.22 (Figure 5c), which was close to the diversity index (1.21) in the center of the modern forest belt (Table 2). In addition, the observed pollen taxa diversity indicated by the Simpson index around Sayram Lake was also very low, reaching a mean of 0.51 (Figure 5d), which was close to the diversity index (0.50) in the center of the modern forest belt (Table 2). Therefore, Sayram Lake was probably in the center of the forest belt during the middle Holocene, and the forest-steppe ecotone probably moved downward accordingly (Figure 6b).

As the late Holocene climate became cold and wet from 4000 yr BP onwards (Figure 5a,b), the conifer forest belt moved downward, and the belts of alpine and sub-alpine meadows and the typical steppe moved down as well (Figure 6c), because the observed pollen taxa diversity indicated by the Shannon-Wiener index around Sayram Lake was relatively high, reaching a mean of 1.36 (Figure 5c), which was close to the diversity index (1.49) of the upper boundary of the modern forest belt (Table 2). In addition, the observed pollen taxa diversity indicated by the Simpson index around Sayram Lake was also relatively high, reaching a mean of 0.65 (Figure 5d), which was close to the diversity index (0.63) of the upper boundary of the modern forest belt (Table 2). Therefore, we infer that Sayram Lake was probably at the upper boundary of the conifer forest during the late Holocene, and the forest-steppe ecotone moved further down and away from Sayram Lake (Figure 6c).

3.4. Pollen Taxa Diversity Change Around Aibi Lake during the Holocene

During the Holocene, the belts of alpine and sub-alpine meadows, conifer forest and typical steppe all contributed to the pollen taxa diversity around Aibi Lake, as the relationships between pollen taxa diversity indicated by the Shannon-Wiener index and the pollen richness from these belts were all positive (Figure 4). Notably, the belt of the conifer forest contributed the most to the pollen taxa diversity around Aibi Lake, according to the result of our linear regression ($R^2 = 0.13$, $p < 0.05$), followed by the alpine and sub-alpine meadows belts ($R^2 = 0.12$, $p < 0.05$) and the typical steppe belt ($R^2 = 0.09$, $p < 0.05$) (Figure 4). This further suggests that pollen taxa diversity around Aibi Lake can reflect the forest belt change during the Holocene.

During the warm and dry periods of the early Holocene from 12,000 to 8000 yr BP (Figure 5a,b), the conifer forest belt moved upward (Figure 6a), and the forest belt contributed less to the low-elevation deposition site of Aibi Lake and no longer dominated its diversity change, as the observed pollen taxa diversity was high, with a value of 1.35 indicated by the Shannon–Wiener index (Figure 5e). As such, we infer that the forest-steppe ecotone moved upward accordingly.

During the middle Holocene from 8000 to 4000 yr BP, when the climate was warm and wet compared to the last stage (Figure 5a,b), the conifer forest belt expanded and dominated (Figure 6b), and the forest belt heavily affected the low-elevation deposition site of Aibi Lake, as the observed pollen taxa diversity was low, with a value of 1.30 indicated by the Shannon-Wiener index, and a value of 0.62 indicated by the Simpson index (Figure 5e,f). Therefore, we infer that the forest-steppe ecotone moved downward accordingly.

As the late Holocene climate became cold and wet from 4000 yr BP onwards (Figure 5a,b), the conifer forest belt moved downward, but its area became smaller with low contribution compared to the last stage, because the observed pollen taxa diversity was high, with a value of 1.46 indicated by the Shannon-Wiener index, and a value of 0.64 indicated by the Simpson index (Figure 5e,f). Therefore, we infer that the forest-steppe ecotone moved further down accordingly.

3.5. Comparison between the Two Lakes

The regression relationships between the pollen taxa diversity calculated using the Shannon-Wiener index and the pollen richness from different vegetation belts showed that pollen taxa diversities around Sayram Lake and Aibi Lake can reflect the forest belt change during the Holocene. When the forest belt expanded, *Picea* pollen dominated, leading to a decrease in the degree of species evenness in the conifer forest belt, because the pollen taxa diversity observed at the two sites was very low.

In general, the pollen taxa diversities around Sayram Lake and Aibi Lake had the same pattern of change in response to the Holocene climate change (Figure 5). In the process of coping with the warm and dry climate during the early Holocene, the forest belt moved upward (Figures 5 and 6a), as Sayram Lake had a high observed pollen taxa diversity (Figures 5 and 6a). In addition, Aibi Lake's pollen taxa diversity was also high, with less contribution from the upward forest belt (Figures 5 and 6a). Therefore, the forest-steppe ecotone moved upward accordingly during the early Holocene. In response to the warm and wet climate during the middle Holocene, the forest belt widened (Figure 6b), as Sayram Lake had very low observed pollen taxa diversity (Figures 5 and 6b). In addition, the *Picea schrenkiana*-dominated forest belt heavily influenced the low-elevation deposition site of Aibi Lake, as the observed pollen taxa diversity was low (Figures 5 and 6b). Hence the forest-steppe ecotone moved downward accordingly during the middle Holocene. In response to the cold and wet climate during the late Holocene, the conifer forest belt moved further downward (Figure 6c), as Sayram Lake had a high observed pollen taxa diversity (Figures 5 and 6c). In addition, Aibi Lake had a high observed pollen taxa diversity, with less contribution from the reduced forest belt (Figures 5 and 6c). Thus, the forest-steppe ecotone moved further down accordingly during the late Holocene.

4. Discussion

Our results show that changes in the forest-steppe ecotone followed the Holocene climate change in the Tien Mountains. As the degree of temperature and moisture changed, the forest belt changed, which further determines the position migration of the forest-steppe ecotone, as supported by the changes in the pollen taxa diversity observed by the deposition sites in the Tien Mountains (Figures 5 and 6). Under the co-influence of solar insolation and westerlies, climate change exhibits a unique pattern in arid central Asia, characterized by warm and dry climates during the early Holocene, warm and wet climates during the middle Holocene, and moderately cold and wet climates during the late Holocene (Figure 5a,b) [16,18,19,29,31]. The forest belt with *Picea* pollen dominance expanded and widened during the middle Holocene, leading to a decrease in the degree of species evenness of the conifer forest belt, as the observed pollen taxa diversity around the two lakes was very low. Specifically, on the northern slope of the Tien Mountains, the mid-elevation humid environment favors the existence of *Picea schrenkiana*, which largely affects mountainous diversity, as supported by our surface pollen results (Figure 3; Table 2). Indeed, this kind of mid-elevation forest distribution is popular in arid and semi-arid mountainous areas in China, which has gained wide attention [5,12,19,32]. Moreover, both the forest belt and the forest-steppe ecotone moved upward during the early Holocene and downward during the late Holocene, leading to a reduced forest belt influence on the sites, as the observed pollen taxa diversities were high (Figure 3; Table 2).

Indeed, Holocene temperature and moisture changes in arid central Asia have affected the vertical migration of vegetation belts and their species composition in mountainous areas [18,19]. The upper limit of the forest belt is controlled by temperature, while the lower boundary is limited by moisture. When the climate is warm and dry, the forest belt moves up along the elevation to acquire more moisture, and vice versa [11,33]. The species composition changes from wet-preferring to drought-tolerant species when climate drying exceeds its physiological limits [5]. Besides this, we found that the forest belt responded to the warm and wet climate through the width of the forest belt, which further made the forest-steppe ecotone move downward accordingly. This corresponding mode was also found in the Holocene change of vegetation belts in the Manas River drainage on the northern slope of the Tien Mountains [14]. Therefore, we emphasize that the migration of the forest-steppe ecotone responded to the Holocene climate change as the forest belt changed, as supported by the change in pollen taxa diversity at the deposition sites.

Although the pollen taxa diversity in arid mountain areas can reflect vegetation belt change (Figures 5 and 6), does a narrowed forest belt lead to a decline in mountain pollen taxa diversity, and further affect the estimation of the forest-steppe ecotone migration? If the width or area of the forest belt becomes narrower, then the pollen taxa diversity observed at the deposition sites is considered to decline. However, we emphasize that if only the width of the forest belt is reduced, while forest species richness and evenness do not change, then the pollen taxa diversity may not necessarily decline. Furthermore, it is worth noting that some taxa in the forest-steppe ecotone may disappear when climate drying exceeds its threshold in arid mountainous areas. Moisture loss may pose the greatest threat to a relatively narrow forest-steppe ecotone, considering the climatic niche space of humidity-sensitive taxa and their limited living space. For example, as one of the diversity hot spots in the world, the narrow vegetation belts of the eastern Andes show great vulnerability to climate warming, as the expected rates of climate change may move the narrowly distributed taxa out of the climate niche space in hundreds of years [34]. More seriously, as global warming intensifies, regional aridity is expected to become more severe and last longer than in the present dry climate in arid regions [3]. Therefore, it seems that future climate warming with the associated lower moisture will probably cause unexpected threats to the taxa with narrow elevational distributions, especially for the taxa in the forest-steppe ecotone in the Tien Mountains. We thus propose that for the protection and maintenance of the current forest-steppe ecotone in mountainous areas, much attention should be paid to humidity-sensitive taxa under the drying climate.

Besides this, pollen taxa diversity may serve as an important proxy for reflecting the migration of the forest-steppe ecotone [4]. However, there are uncertainties, mainly due to the following aspects. Firstly, the limitation of pollen indicators—potential uncertainties come from differences in the pollen productivity and dispersal ability among various species [4,35–37], which may not present a one-to-one dynamic of vegetation belts and richness of fossil pollen. Secondly, the impact of vegetation structure—for wind-transported pollens, their ability to spread in the woodland is much worse than that in open land. This not only affects the amount of pollen that can be observed at the deposition site, but also affects our ability to recover vegetation using pollen data [38]. Thirdly, different pollen types have different representativeness—if a community with the same species composition has different coverage, the yield of pollen will also be different. When the number of statistical grains is constant, the low-representative species may not be counted, which will affect the estimation of pollen taxa diversity and in future will affect the estimation of the migration of the forest-steppe ecotone [38]. Because this study is mainly based on the Shannon-Wiener index and the Simpson index for calculating the pollen taxa diversity, which takes into account both richness and uniformity, taxa with very low pollen numbers or percentages have little effect on the results.

5. Conclusions

We found that the forest-steppe migration generally followed Holocene climate change in the Tien Mountains. Specifically, the forest belt where *Picea schrenkiana* dominates has a very low pollen taxa diversity, characterized by high richness and low evenness, which plays a key role in the mountainous diversity. By detecting the diversity change of the deposition sites, we found that in the process of coping with warm and dry climate during the early Holocene from 12,000 to 8000 yr BP, the forest belt moved upward, causing a reduced forest belt influence on, or contribution to, the low-elevation deposition sites, as the observed diversities around the two lakes were high. In this case, the forest-steppe ecotone moved upward accordingly. During the warm and wet middle Holocene from 8000 to 4000 yr BP, the forest belt with low taxa evenness expanded and widened, as the observed diversities around the two lakes were very low. As such, the forest-steppe ecotone moved downward accordingly. As the late Holocene became cold and wet from 4000 yr BP onwards, the forest belt moved downward, resulting in a reduced forest belt influence on or contribution to the sites, leading to the high diversities observed.

Thus the forest-steppe ecotone moved further down. Moisture loss may pose the greatest threat to the narrow forest-steppe ecotone, considering the climatic niche space and the limited living space for humidity-sensitive taxa. This study highlights that temperature and moisture co-influence the forest belt change, which further determines the migration of the forest-steppe ecotone.

Author Contributions: H.L. and Y.C. proposed the idea and designed the study. Y.C. analyzed the data and wrote the first draft. Y.C., H.L., H.W., Q.H., Y.H., K.D. and Z.D. discussed the results and contributed to improving the manuscript. All authors have read and agreed to the published version of the manuscript.

Funding: This research was funded by the National Natural Science Foundation of China, Grant Number 41790422, 41901092. It was also funded by the Fundamental Research Funds for the Central Universities of China, Grant Number GK202003069.

Conflicts of Interest: The authors declare no conflict of interest.

References

1. Bailey, R.M. Spatial and temporal signatures of fragility and threshold proximity in modelled semi-arid vegetation. *Proc. R. Soc. B* **2010**, *278*, 1064–1071. [CrossRef]

2. Liu, H.Y.; Park Williams, A.; Allen, C.D.; Guo, D.; Wu, X.; Anenkhonov, O.A.; Badmaeva, N.K.; Liang, E.; Sandanov, D.V.; Yin, Y.; et al. Rapid warming accelerates tree growth decline in semi-arid forests of inner Asia. *Glob. Chang. Biol.* **2013**, *19*, 2500–2510. [CrossRef]

3. Huang, J.; Yu, H.; Guan, X.; Wang, G.; Guo, R. Accelerated dryland expansion under climate change. *Nat. Clim. Chang.* **2016**, *6*, 166–171. [CrossRef]

4. Weng, C.; Hooghiemstra, H.; Duivenvoorden, J.F. Response of pollen diversity to the climate-driven altitudinal shift of vegetation in the Colombian Andes. *Phil. Trans. R. Soc. B* **2007**, *362*, 253–262. [CrossRef]

5. Cheng, Y.; Liu, H.; Dong, Z.; Duan, K.; Wang, H.; Han, Y. East Asian summer monsoon and topography co-determine the Holocene migration of forest-steppe ecotone in northern China. *Glob. Planet. Chang.* **2020**, *187*, 103135. [CrossRef]

6. Franco, A.M.; Hill, J.K.; Kitschke, C.; Collingham, Y.C.; Roy, D.B.; Fox, R.; Huntley, B.; Thomas, C.D. Impacts of climate warming and habitat loss on extinctions at species' low-latitude range boundaries. *Glob. Chang. Biol.* **2006**, *12*, 1545–1553. [CrossRef]

7. La Sorte, F.A.; Jetz, W. Projected range contractions of montane plant species diversity under global warming. *Proc. R. Soc. B* **2010**, *277*, 3401–3410.

8. Wu, J.; Zhang, S.; Jiang, Y. *Botanical Geography*; China Higher Education Press: Beijing, China, 1995.

9. Erdos, L.; Ambarlı, D.; Anenkhonov, O.A.; Bátori, Z.; Cserhalmi, D.; Kiss, M.; Naqinezhad, A.; Kröel-Dulay, G.; Liu, H.; Magnes, M.; et al. The edge of two worlds: A new review and synthesis on Eurasian forest-steppes. *Appl. Veg. Sci.* **2018**, *21*, 345–362. [CrossRef]

10. Dai, J.; Liu, H.; Xu, C.; Qi, Y.; Zhu, X.; Zhou, M.; Wu, Y.; Liu, B. Divergent hydraulic strategies explain the interspecific associations of co-occurring trees in forest–steppe ecotone. *Forests* **2020**, *11*, 942.

11. Cheng, Y.; Liu, H.; Wang, H.; Piao, S.; Yin, Y.; Ciais, P.; Gao, Y.; Wu, X.; Luo, Y.; Zhang, C.; et al. Contrasting effects of winter and summer climate on alpine timberline evolution in monsoon-dominated East Asia. *Quat. Sci. Rev.* **2017**, *169*, 278–287. [CrossRef]

12. Huang, X.Z.; Peng, W.; Rudaya, N.; Grimm, E.C.; Chen, X.; Cao, X.; Chen, F.; Zhang, J.; Pan, X.; Liu, S.; et al. Holocene vegetation and climate dynamics in the Altai Mountains and surrounding areas. *Geophys. Res. Lett.* **2018**, *45*, 6628–6636. [CrossRef]

13. Zhao, Y.; Tzedakis, P.C.; Li, Q.; Qin, F.; Cui, Q.; Liang, C.; Zhao, H.; Birks, H.J.B.; Liu, Y.; Zhang, Z.; et al. Evolution of vegetation and climate variability on the Tibetan Plateau over the past 1.74 million years. *Sci. Adv.* **2020**, *6*, eaay6193. [CrossRef] [PubMed]

14. Ji, Z.K.; Liu, H.Y. Shifting of vertical vegetation zones in Manas River drainage on northern slope of Tianshan Mountains since the Late Glacier. *J. Palaeogeo.* **2009**, *11*, 534–541.

15. Gottfried, M.; Pauli, H.; Futschik, A.; Akhalkatsi, M.; Barančok, P.; Alonso, J.L.B.; Coldea, G.; Dick, J.; Erschbamer, B.; Calzado, M.R.F.; et al. Continent-wide response of mountain vegetation to climate change. *Nat. Clim. Chang.* **2012**, *2*, 111–115. [CrossRef]

16. Chen, F.H.; Yu, Z.; Yang, M.; Ito, E.; Wang, S.; Madsen, D.B.; Huang, X.; Zhao, Y.; Sato, T.; Birks, H.J.B.; et al. Holocene moisture evolution in arid central Asia and its out-of-phase relationship with Asian monsoon history. *Quat. Sci. Rev.* **2008**, *27*, 351–364. [CrossRef]

17. Long, H.; Shen, J.; Chen, J.; Tsukamoto, S.; Yang, L.; Cheng, H.; Frechen, M. Holocene moisture variations over the arid central Asia revealed by a comprehensive sand-dune record from the central Tian Shan, NW China. *Quat. Sci. Rev.* **2017**, *174*, 13–32. [CrossRef]

18. Zhang, D.; Chen, X.; Li, Y.; Wang, W.; Sun, A.; Yang, Y.; Ran, M.; Feng, Z. Response of vegetation to Holocene evolution of westerlies in the Asian Central Arid Zone. *Quat. Sci. Rev.* **2020**, *229*, 106138. [CrossRef]

19. Huang, X.Z.; Chen, C.Z.; Jia, W.N.; An, C.B.; Zhou, A.F.; Zhang, J.W.; Grimm, E.C.; Jin, M.; Xia, D.; Chen, F. Vegetation and climate history reconstructed from an alpine lake in central Tienshan Mountains since 8.5 ka BP. *Palaeogeogr. Palaeoecol.* **2015**, *432*, 36–48. [CrossRef]

20. Sun, X.; Du Naiqiu, W.C. Paleovegetation and paleoenvironment of Manasi Lake, Xinjiang, NW China during the last 14,000 years. *Quat. Sci.* **1994**, *3*, 239–248.

21. Li, J.; Gou, X.; Cook, E.R.; Chen, F. Tree-ring based drought reconstruction for the central Tien Shan area in northwest China. *Geophys. Res. Lett.* **2006**, *33*, L07715. [CrossRef]

22. Jiang, Q.F.; Ji, J.; Shen, J.; Matsumoto, R.; Tong, G.; Qian, P.; Yan, D.; Ren, X. Holocene vegetational and climatic variation in westerly-dominated areas of Central Asia inferred from the Sayram Lake in northern Xinjiang, China. *Sci. China Earth Sci.* **2013**, *56*, 339–353. [CrossRef]

23. Wang, W.; Feng, Z.; Ran, M.; Zhang, C. Holocene climate and vegetation changes inferred from pollen records of Lake Aibi, northern Xinjiang, China: A potential contribution to understanding of Holocene climate pattern in East-central Asia. *Quat. Int.* **2013**, *311*, 54–62. [CrossRef]

24. Chen, X.; Luo, G.; Xia, J.; Zhou, K.; Lou, S.; Ye, M. Ecological response to the climate change on the northern slope of the Tianshan Mountains in Xinjiang. *Sci. China Earth Sci.* **2005**, *48*, 765–777. [CrossRef]

25. Moore, P.D.; Webb, J.A.; Collison, M.E. *Pollen Analysis*; Blackwell Scientific Publications: Oxford, UK, 1991.

26. Cheng, Y.; Liu, H.; Wang, H.; Hao, Q. Differentiated climate-driven Holocene biome migration in western and eastern China as mediated by topography. *Earth Sci. Rev.* **2018**, *182*, 174–185. [CrossRef]

27. Gophen, M. Temperature Impact on the *Shannon-Wiener Plant* species diversity index (BDI) of Zooplankton in Lake Kinneret (Israel). *Open J. Mod. Hydrol.* **2018**, *8*, 39–49. [CrossRef]

28. Morris, E.K.; Caruso, T.; Buscot, F.; Fischer, M.; Hancock, C.; Maier, T.S.; Meiners, T.; Müller, C.; Obermaier, E.; Socher, S.A.; et al. Choosing and using diversity indices: Insights for ecological applications from the German biodiversity exploratories. *Ecol. Evol.* **2014**, *18*, 3514–3524. [CrossRef]

29. Marcott, S.A.; Shakun, J.D.; Clark, P.U.; Mix, A.C. A reconstruction of regional and global temperature for the past 11,300 years. *Science* **2013**, *339*, 1198–1201. [CrossRef]

30. Chen, F.; Wu, D.; Chen, J.; Zhou, A.; Yu, J.; Shen, J.; Wang, S.; Huang, X. Holocene moisture and East Asian summer monsoon evolution in the northeastern Tibetan Plateau recorded by Lake Qinghai and its environs: A review of conflicting proxies. *Quat. Sci. Rev.* **2016**, *154*, 111–129. [CrossRef]

31. Zhao, Y.; Yu, Z. Vegetation response to Holocene climate change in East Asian monsoon-margin region. *Earth Sci. Rev.* **2012**, *113*, 1–10. [CrossRef]

32. Hao, Q.; Liu, H.; Liu, X. Pollen-detected altitudinal migration of forests during the Holocene in the mountainous forest–steppe ecotone in northern China. *Palaeogeogr. Palaeoecol.* **2016**, *446*, 70–77. [CrossRef]

33. Liu, H.Y.; Yin, Y.; Zhu, J.; Zhao, F.; Wang, H. How did the forest respond to Holocene climate drying at the forest-steppe ecotone in northern China? *Quatern. Int.* **2010**, *227*, 46–52. [CrossRef]

34. Bush, M.B.; Silman, M.R.; Urrego, D.H. 48,000 years of climate and forest change in a plant species diversity hot spot. *Science* **2004**, *303*, 827–829. [CrossRef] [PubMed]

35. Sugita, S. Theory of quantitative reconstruction of vegetation II: All you need is LOVE. *Holocene* **2007**, *17*, 243–257. [CrossRef]

36. Sugita, S. Theory of quantitative reconstruction of vegetation I: Pollen from large sites REVEALS regional vegetation composition. *Holocene* **2007**, *17*, 229–241. [CrossRef]
37. Han, Y.; Liu, H.; Hao, Q.; Liu, X.; Guo, W.; Shangguan, H. More reliable pollen productivity estimates and relative source area of pollen in a forest-steppe ecotone with improved vegetation survey. *Holocene* **2017**, *27*, 1567–1577. [CrossRef]
38. Liu, H.Y. *Quaternary Ecology and Global Change*; China Science Press: Beijing, China, 2002.

Publisher's Note: MDPI stays neutral with regard to jurisdictional claims in published maps and institutional affiliations.

Article

Divergent Hydraulic Strategies Explain the Interspecific Associations of Co-Occurring Trees in Forest–Steppe Ecotone

Jingyu Dai [1], Hongyan Liu [1,*], Chongyang Xu [1], Yang Qi [1], Xinrong Zhu [1], Mei Zhou [2], Bingbing Liu [2,3] and Yiheng Wu [2]

[1] College of Urban and Environmental Science and MOE Laboratory for Earth Surface Processes, Peking University, Beijing 100871, China; daijingyu@pku.edu.cn (J.D.); xchongyang1008@126.com (C.X.); qi_yang@pku.edu.cn (Y.Q.); zhuxinrong@pku.edu.cn (X.Z.)
[2] College of Forestry, Inner Mongolia Agricultural University, Inner Mongolia 010000, China; dxal528@aliyun.com (M.Z.); bingbing-l@caf.ac.cn (B.L.); uykhan@163.com (Y.W.)
[3] Institute of Forest Ecology, Environment and Protection, Chinese Academy of Forestry, Beijing 100091, China
* Correspondence: lhy@urban.pku.edu.cn

Received: 11 July 2020; Accepted: 27 August 2020; Published: 28 August 2020

Abstract: *Research Highlights:* Answering how tree hydraulic strategies explain the interspecific associations of co-occurring trees in forest–steppe ecotone is an approach to link plant physiology to forest dynamics, and is helpful to predict forest composition and function changes with climate change. *Background* and *Objectives:* The forest–steppe ecotone—the driest edges of forest distribution—is continuously threatened by climate change. To predict the forest dynamics here, it is crucial to document the interspecific associations among existing trees and their potential physiological drivers. *Materials* and *Methods:* Forest–steppe ecotone is composed of forest and grassland patches in a mosaic pattern. We executed two years of complete quadrat surveys in a permanent forest plot in the ecotone in northern China, calculated the interspecific association among five main tree species and analyzed their hydraulic strategies, which are presented by combining leaf-specific hydraulic conductivity (K_l) and important thresholds on the stem-vulnerability curves. *Results:* No intensive competition was suggested among the co-occurring species, which can be explained by their divergent hydraulic strategies. The negative associations among *Populus davidiana* Dode and *Betula platyphylla* Suk., and *P. davidiana* and *Betula dahurica* Pall. can be explained as the result of their similar hydraulic strategies. *Tilia mongolica* Maxim. got a strong population development with its effective and safe hydraulic strategy. Generally, hydraulic-strategy differences can explain about 40% variations in interspecific association of species pairs. Oppositely, species sensitivity to early stages of drought is convergent in the forest. *Conclusions:* The divergent hydraulic strategies can partly explain the interspecific associations among tree species in forest–steppe ecotone and may be an important key for semiarid forests to keep stable. The convergent sensitivity to early stages of drought and the suckering regeneration strategy are also important for trees to survival. Our work revealing the physiological mechanism of forest compositions is a timely supplement to forest–steppe ecotone vegetation prediction.

Keywords: drought tolerance; forest–steppe ecotone; hydraulic strategy; hydraulic trait; interspecific association; interspecific relationships; species co-occurrence; semiarid forests

---

1. Introduction

The forest–steppe ecotone is located at the driest edges of forest distribution and is continually threatened by the increasing temperature and drought [1–4]. Forest patches and grassland patches distribute in a mosaic pattern in the forest–steppe ecotone [4]. To answer whether the forests are stable in the forest–steppe ecotone, it is helpful to explore what interspecific associations exist among tree species and what physiological mechanisms are driving them, as the interspecific associations of co-occurring trees will determine the forest composition variations [5].

Interspecific association—referring to the relationships and interactions among co-occurring species [6]—has been reported to be driven by many factors, including the forest succession process [5], the heterogeneity of microenvironment [7], etc. In the forest–steppe ecotone—the driest edges of forest distribution—water deficit is the most dominant environment stress affecting forest dynamics and functions [2,8], instead of reported interspecific association drivers. Moreover, water deficit here has been predicted to be exacerbated in the future [2,8]. For this reason, it is crucial to explore the importance and role of water deficit in shaping the vegetation community in the forest–steppe ecotone.

Facing the aggravated water deficit, plant hydraulic strategies balancing hydraulic safety and water-transport efficiency are believed to determine tree drought responses [9–11]. The water-transport efficiency of trees has direct impacts on the carbon uptake and productivity of trees, while xylem resistance to cavitation determines the safety of individuals when extreme drought events occur [11–13]. The safety and efficiency of trees generally occur as a tradeoff. A larger conduit diameter provides higher hydraulic efficiency as well as higher vulnerability to drought-induced embolism in the xylem, and vice versa [14,15]. The hydraulic safety–efficiency combination of trees can provide a way to map hydraulic strategy of each tree species, to describe the differences among the hydraulic strategies of co-occurring species and to judge the vulnerability of trees in the forest–steppe ecotone under ongoing climate change [16,17].

If water deficit plays a considerable role in forest interspecific associations, divergent hydraulic strategies should differentiate the water-use strategies of species and reduce interspecific water competitions [18], thus benefitting species co-occurrence. Further, we can expect the differences of species hydraulic strategies to be an effective predictor of interspecific relationships, i.e., the hydraulic strategies of co-existent species to be divergent, while the hydraulic strategies of mutually exclusive species to be convergent. Meanwhile, the species with both higher hydraulic safety and efficiency should have higher potential to be dominant in the community. Forest–steppe ecotone has all the ten genera of tree species distributing in semi-humid forests in northern China, even with a mean annual precipitation of about 300 mm [19]. If the diversity of tree hydraulic strategies can provide partitioning of the water resource, it may be the reason for forest–steppe ecotone to maintain the high tree species diversity. In general, we developed two hypotheses in this article, 1) there is a positive correlation between interspecific association and distance in species hydraulic strategy, and 2) species with both high hydraulic efficiency and safety may play a dominant role in the forest, or have experienced a rapid population development over the years.

To verify the hypotheses, we carried out a two-year vegetation survey and a series of hydraulic measurements in a permanent plot in the Saihan Wula National Nature Reserve, Inner Mongolia, China. In this work, hydraulic efficiency is represented by leaf-specific hydraulic conductivity (K_l), while hydraulic safety is represented by a series of important thresholds on the stem-vulnerability curves, stem water potential at a 50% loss of stem conductivity (P_{50}) especially. The K_l and P_{50} are combined to show the hydraulic strategy of each species.

2. Materials and Methods

2.1. Site Description and Sampling Design

Field work was conducted within a 6-ha permanent plot in a temperate deciduous broadleaved forest in the Saihan Wula National Nature Reserve in the southern Da Hinggan Mountains, Inner Mongolia, China (Figure 1a,b, 44°12′ N, 118°44′ E). The plot is located in semiarid forest–steppe ecotone, with a mean annual temperature of 2 °C and a mean annual precipitation of 358 mm. The wet season is from May to September. The elevation of the plot is approximately 1175–1355 m ASL [20].

The studied forest is located on the north side of the mountain, while the steppe lies on the south (Figure 1d). Severe drought-induced forest mortality occurred during 2011–2012, followed by a huge amount of regeneration. Tree density increased from 1825.17 ha^{-1} in 2012 to 3812.17 ha^{-1} in 2015. No grazing, agricultural abandonment, forest management or fire interference have been found inside or around the plot since 1997.

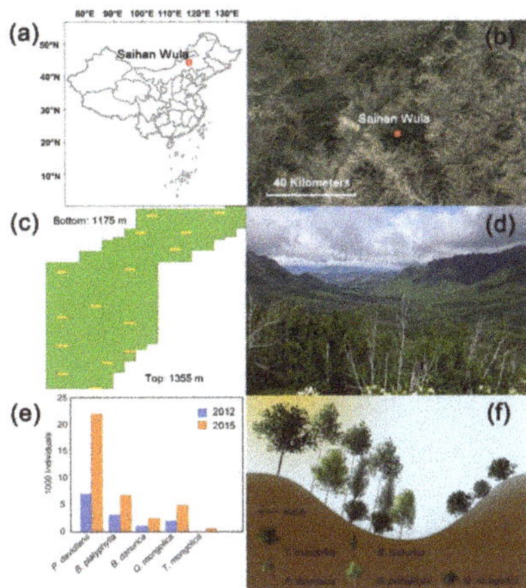

Figure 1. Study area location and site characteristics. (**a**) Location of the study area in China; (**b**) landscape of semiarid forest–steppe ecotone in which the study area lies; (**c**) sketch of the plot (shown in green polygon) and quadrats containing sampled trees for hydraulic architecture measurements (shown as yellow squares). Each quadrat has an area of 100 m^2; (**d**) landscape photography of the study area showing the mosaic distribution of forest (dark green patches) and grassland (light green patches); (**e**) amount of each tree species in 2012 (shown in blue) and 2015 (shown in orange); (**f**) schematic picture showing the forest structure within the plot.

Two comprehensive plot surveys were carried out in the whole permanent plot in 2012 and 2015, respectively. The permanent plot was divided into 608 contiguous 10 m × 10 m quadrats. Basic information of all trees with diameter at breast height >1 cm was carefully recorded, including species, tree height, diameter at breast height and the located quadrat identification number. Only main species were taken into further analysis, with the proportion in the total tree number over 1%, which are *Populus davidiana* Dode, *Betula platyphylla* Suk., *Betula dahurica* Pall., *Quercus mongolica* Fisch. ex Ledeb and *Tilia mongolica* Maxim. (Figure 1e,f). *P. davidiana*, *B. platyphylla* and *B. dahurica* are early successional species, while *Q. mongolica* and *T. mongolica* are late successional species. *P. davidiana* was the most dominant tree species in the forests, accounting for the largest amount of the tree mortality and regeneration during and after the extensive drought events in 2011–2012 (Figure 1e), with a mortality ratio of over 60% [21].

Hydraulic traits were measured for five trees per species at different elevations (Figure 1c). The maximal vessel lengths of each species are: *P. davidiana*: 11 cm, *B. platyphylla*: 14 cm, *B. dahurica*: 16 cm, *Q. mongolica*: 86 cm and *T. mongolica*: 20 cm.

2.2. Interspecific Association Quantification

The interspecific association of five species were quantified with the plot survey data. The variance ratio (VR) was used to test the overall association among species. If VR = 1, species are independent from each other. A VR >1 indicates positive correlation, while a VR <1 indicates negative correlation. Statistics W and χ^2 test were used to test the significance of the difference between VR value and 1, with a $\chi^2_{(0.05, N)} < W < \chi^2_{(0.95, N)}$ indicating statistically insignificant difference between VR value and 1, which suggests the original hypothesis is true. The formulas are listed below [6]:

$$VR = \frac{S_T^2}{\delta_T^2} = \frac{\frac{1}{N} \sum_{j=1}^{N} (T_j - t)^2}{\sum_{i=1}^{S} (P_i \times (1 - P_i))} \left(P_i = \frac{n_i}{N} \right) \tag{1}$$

$$W = VR \times N \tag{2}$$

where S represents the total species number, N represents the total quadrat number, T_j represents the number of species occurring in the quadrat j, t represents the average number of species occurring in each quadrat, while P_i represents the percentage of the quadrats with the species i occurring in them.

Then, based on 2×2 contingency table among species, we used the Yates correlation coefficient, Ochiai index and principal component analysis (PCA) to reveal the interspecific association between each pair of species. First, Yates correlation coefficient was processed to test whether the associations of species pairs existed with the predetermined probability. In Yates test, index $V > 0$ indicates a positive correlation between the two species, while $V < 0$ indicates a negative correlation between species. χ^2 is used to show the significance of the correlation. If $\chi^2 < 3.841$, the two species are independently distributed, if $3.841 \leq \chi^2 < 6.635$, significant association is attached, and if $\chi^2 > 6.635$, the association between two species is highly significant [6]. Second, Ochiai index was used to measure the degree of the associations. Ochiai index distributed between 0 and 1. A higher OI indicates higher possibility for two species to occur in the same quadrat. The formulas of Yates correlation and Ochiai index are listed below [5,6]:

$$V = ad - bc \tag{3}$$

$$\chi^2 = \frac{n(|ad - bc| - 0.5n)^2}{(a+b)(a+c)(b+d)(c+d)} \tag{4}$$

$$OI = \frac{a}{\sqrt{(a+b)(a+c)}} \tag{5}$$

where n represents the total quadrat number, a represents the number of quadrats with both two species, b and c represent the number of quadrats with only one species, while d represents the number of quadrats with neither of the two species. Finally, PCA analysis was adopted to show the relationship among species and quadrats at the community level.

2.3. Hydraulic Traits Measurement

Long branches over 50 cm with diameters of approximately 5 to 8 mm were cut by clippers, and all the leaves on the cut branch were retained. The branch samples were placed in black plastic bags, and the cross sections were immediately wrapped in moist tissue in the field to reduce moisture loss. The branch samples were cut twice into 27.4-cm length segments under water in the laboratory [22], removing the excess length from both ends, ensuring shaving off more than 5 cm from each end [22,23] and retaining a straight segment with as few embranchments as possible.

Maximum hydraulic conductivity (K_{max}, 10^{-4} kg m MPa^{-1} s^{-1}) was measured for each sample by flushing the stem segments with the 20-mM KCl solution at 0.1 MPa for 30 min and measuring the flow rate of ultrafiltered water passing through the segment under pressure difference of around 0.05 MPa [15,24]. All the leaves above the sampled branches were scanned using a portable scanner (Founder MobileOffice Z6, Beijing, China) and the total leaf area was calculated using MATLAB R2014a (The MathWorks, Natick, MA, USA). Leaf specific conductivity (K_l, 10^{-4} kg m^{-1} s^{-1} MPa^{-1}) was then determined as the ratio of K_{max} and the total leaf area of the branches. K_l is a synthetic proxy for the balance between the water demand of leaves and the hydraulic transport efficiency of stems [14]. All the measurements were made within two days after sampling.

The vulnerability of the stems to drought-induced cavitation, i.e., the stem-vulnerability curve, was estimated using the centrifugal force method [25] with a superspeed centrifuge (CTK150R, XiangYi Centrifuge Instrument, Changsha, China). The segment ends were kept immersed in water during spinning [25]. The K_{max} of the stem segments was set as the starting point of the curve. Then, the hydraulic conductance (K_h, 10^{-4} kg m MPa^{-1} s^{-1}) were measured after a series of stepwise increasing xylem tensions caused by different spinning speeds. The vulnerability curve was finally plotted as the variation in the percentage loss of hydraulic conductivity (PLC) with increasing xylem tension. The PLC of the stems was calculated as:

$$PLC = [(K_{max} - K_h)/K_{max}] \tag{6}$$

From the vulnerability curves, we can extract some important water potential thresholds, which mark declines in stomatal conductance, hydraulic conductivity and potential for survival. Here, we choose stem water potential at 12% (P_{12}), 50% (P_{50}) and 88% (P_{88}) loss of stem conductivity as the proxies for the early declines in water supply influencing leaf stomatal activities, gas exchange and carbon uptake, the maximum water stress trees can bear naturally and the threshold of irreversible xylem functional damage, respectively [26].

2.4. Statistical Analysis

The stem-vulnerability curves were fitted with a four-parameter Weibull model using SigmaPlot v14.0 (Systat Software, Inc., San Jose, CA, USA). Variance analysis was used to show the differences among the hydraulic traits of different species, followed by least significant difference (LSD) post hoc tests. To reveal the relationship between the interspecific association pattern and the hydraulic-strategy differences among species, we calculated the hydraulic differences between each pair of species with K_l and P_{50}, including Euclidean, Manhattan, Canberra, Minkowski and maximum distance. If all of the distances show similar relationships with interspecific association indices, we can accept the hypothesis that linkages exist between hydraulic-strategy differences and interspecific associations. Correlation tests and linear regression were used further for the hydraulic differences and the interspecific association indices, V_{2012}, V_{2015}, OI_{2012} and OI_{2015} to map the relationship between them. All the statistical analyses were performed in R software (R Development Core Team, 2009), while figures were drawn with R software and SigmaPlot v14.0.

3. Results

3.1. Interspecific Association among Species

The overall species association measured by the VR test revealed no significant correlation within the forest community in either 2012 or 2015 (in 2012, VR = 1.10 > 1, W = 569.10, $\chi^2_{(0.95, N)}$ = 569.95, $\chi^2_{(0.05, N)}$ = 464.32; in 2015, VR = 1.03 > 1, W = 604.23, $\chi^2_{(0.95, N)}$ = 646.57, $\chi^2_{(0.05, N)}$ = 533.71).

The association measurements between each pair of species revealed generally negative correlations between *P. davidiana* and co-occurring species (Table 1). In 2012, there was no significant correlation between *P. davidiana* and *T. mongolica*, while the relationship between *P. davidiana* and *B. dahurica*, *B. platyphylla* and *Q. mongolica* were all negative. In 2015, the relationship between *P. davidiana* and all the co-occurring species became negatively correlated.

Table 1. Yates correlations and Ochiai indices showing interspecific associations among main species in the forest in 2012 and 2015.

Species		Yates Correlation				Ochiai Index	
		V_{2012}	χ^2_{2012}	V_{2015}	χ^2_{2015}	OI_{2012}	OI_{2015}
P. davidiana	*B. platyphylla*	−12,384	31.80	−15,998	34.44	0.39	0.38
P. davidiana	*B. dahurica*	−4816	2.74	−11,818	12.74	0.52	0.50
P. davidiana	*Q. mongolica*	−8944	10.07	−1083	0.07	0.56	0.70
P. davidiana	*T. mongolica*	172	0.13	−8227	14.45	0.05	0.21
B. platyphylla	*B. dahurica*	11,592	24.83	15,661	27.45	0.68	0.67
B. platyphylla	*Q. mongolica*	9036	15.09	2620	0.73	0.73	0.75
B. platyphylla	*T. mongolica*	−420	0.65	2353	1.18	0.00	0.33
B. dahurica	*Q. mongolica*	17,693	36.26	15,246	20.70	0.64	0.64
B. dahurica	*T. mongolica*	−239	0.01	8220	12.01	0.00	0.33
Q. mongolica	*T. mongolica*	223	0.02	−2756	1.34	0.06	0.26

The interspecies association among *B. dahurica*, *B. platyphylla* and *Q. mongolica* in both years were strong and significantly positive ($V > 0$, $\chi^2 > 6.635$ for all associations except the one between *B. platyphylla* and *Q. mongolica* in 2015; $OI > 0.6$), showing the distribution of the three species tend to be convergent (Table 1).

A strong increase of *T. mongolica* could be detected from 2012 to 2015. The total number of *T. mongolica* in plot increased from three in 2012 to 486 in 2015 (Figure 1e), and the percent of quadrats with *T. mongolica* increased from 0.2% in 2012 to 12.0% in 2015. Consequently, the association between *T. mongolica* and co-occurring species had prominently been increased (Table 1). In 2012, none of the association between *T. mongolica* and other species were significant ($\chi^2 < 3.841$ in Yates correlation; $OI < 0.1$), while in 2015, *T. mongolica* showed extremely significant positive correlation with *B. dahurica* and negative correlation with *P. davidiana* ($\chi^2 > 6.635$). For Ochiai index, though the association between *T. mongolica* and other species were both weak, OI value in 2015 were generally higher than OI in 2012.

Principle component analysis (PCA) integrated the interspecific associations within the forest in these two years (Figure 2). The first two components reflected 52.53% and 52.42% of the total variabilities in 2012 and 2015, respectively. *B. platyphylla* and *B. dahurica* located nearby the first component at the opposite of *P. davidiana* in both years. *Q. mongolica* located close to the first component in 2012, while close to the second component in 2015, indicating *Q. mongolica* being increasing decoupled with *P. davidiana*, *B. platyphylla* and *B. dahurica* over the years. The location of *T. mongolica* in 2015 was inversed to it in 2012 and be closer to the first component than 2012, indicating a closer association emerged over the years. Quadrats at the bottom of the mountain located generally at the top–left in PCA figures, while quadrats at the top of the mountain usually distributed at the bottom–right in figures, indicating *P. davidiana* to occupy more proportion at the bottom of the mountain, while *T. mongolica* and *B. platyphylla* occupying more proportion at the top of the mountain.

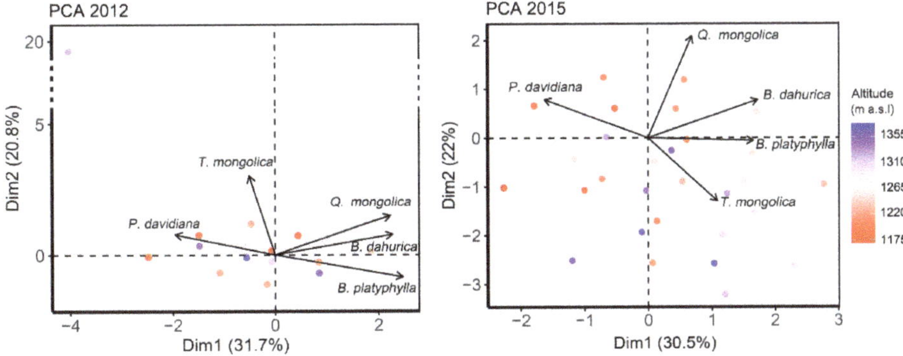

Figure 2. Principle component analysis (PCA) showing the interspecific associations among main species in the forest in 2012 and 2015. Arrows show the load of each species on each principle components. Dots indicate the PCA scores for the quadrats. Color of the dots represents the location of the quadrats along the altitude.

3.2. Hydraulic Safety and Efficiency Differences among Species

There were significant differences in K_l between species (Figure 3). The K_l value of *B. dahurica*, *B. platyphylla*, *T. mongolica* and *Q. mongolica* increased significantly in sequence ($p < 0.05$), ranging from 1.1×10^{-4} to 11.2×10^{-4} kg m MPa^{-1} s^{-1}. There were no significant differences between the K_l value of *P. davidiana* and *B. dahurica*, nor between *P. davidiana* and *B. platyphylla*.

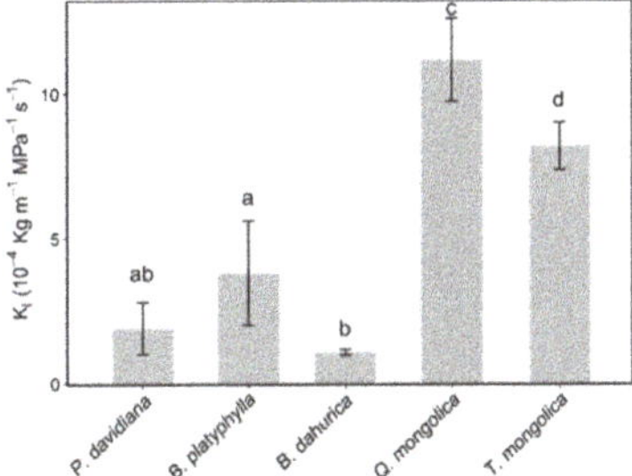

Figure 3. Leaf-specific hydraulic conductivity (K_l) of the co-occurring species are shown in the bar figure. Error bars show ± 1 SE. Different characters show the significant differences according to a variance analysis ($p < 0.05$).

The stem-vulnerability curves showed generally convergent sensitivity to early stages of drought, but different stem cavitation resistance abilities among co-occurring tree species (Figure 4). All the species except *Q. mongolica* had the same P_{12}. P_{12} of *Q. mongolica*, -1.24 MPa, was significantly lower than *B. platyphylla*, *B. dahurica* and *T. mongolica*, indicating *Q. mongolica* to be more sensitive to the early stages of drought. Though some species pairs did not show significant differences, the P_{50} and P_{88} values indicated the stem cavitation resistance of species to be generally ordered as: *T. mongolica* = *B. platyphylla* > *P. davidiana* > *B. dahurica* > *Q. mongolica*.

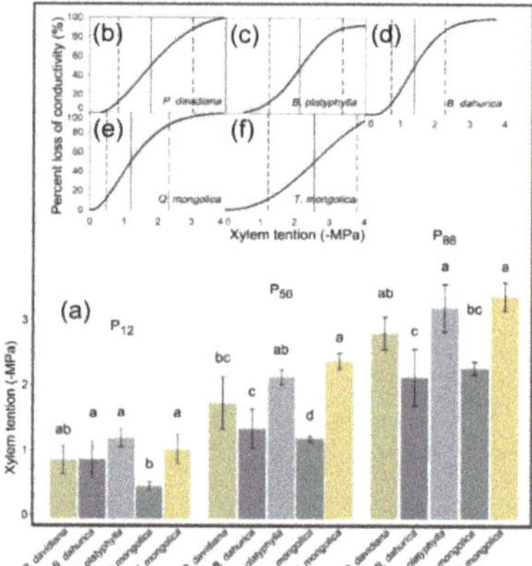

Figure 4. (**a**) Comparison of the water potential of stems at 12%, 50% and 88% loss of stem conductivity (P_{12}, P_{50} and P_{88}, respectively) of each species. The different characters indicate the significant differences among the thresholds of each species based on variance analysis; (**b**–**f**) Vulnerability curves are shown as the dynamic of percent loss of stem hydraulic conductivity in response to centrifugal-force xylem tension for each tree species. Gray curves are fitted by each sample, while the black curves are fitted by the mean values. For each species, P_{12}, P_{50} and P_{88} are shown as the single-dashed, solid and double-dashed vertical lines, respectively.

The interspecific differences on hydraulic strategies between safety and efficiency can be integrated as Figure 5 when representing xylem hydraulic safety by P_{50}. Generally speaking, the hydraulic strategies of different species were divergent. Hydraulic strategy of *P. davidiana* has much overlap with *B. platyphylla* and *B. dahurica*, though the strategies of the latter two were different. *Q. mongolica* located at the right bottom corner, suggesting the xylem have high hydraulic conductance but poor ability for drought resistance. *T. mongolica* lied in the right top corner of figure, suggesting *T. mongolica* have both high hydraulic conductance and high threshold for cavitation. The same pattern could be found if the xylem hydraulic safety were represented by P_{88} (Figure S1).

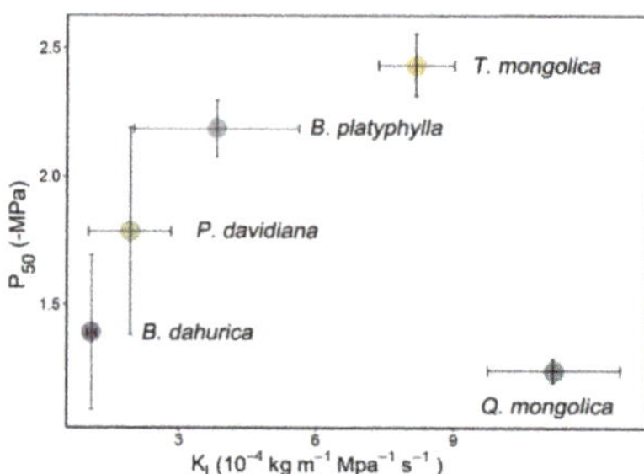

Figure 5. Relationship between stem hydraulic transportation efficiency (as measured by leaf-specific hydraulic conductivity, K_l) and safety (as measured by the water potential of stems at 50% loss of stem conductivity, P_{50}) of all species. Error bars show ± 1 SE.

3.3. The Relationship between Interspecific Association and Species Hydraulic Strategy

The correlation between interspecific association indices and distances between species pairs calculated by hydraulic traits, K_l and P_{50}, were generally positive, but not significant. The one between Canberra distance and V value in Yates correlation in 2015 was significant (Figure S2). By linear regression, V_{2015} could be predicted by Canberra distance with the model

$V_{2015} = -19{,}283 + 28{,}794 \times$ Canberra distance (Figure 6, $p = 0.05$, $R^2 = 0.39$). About 40% of the variance of interspecific association could be explained by the interspecific differences on hydraulic strategies in the forest we studied (Figure 6).

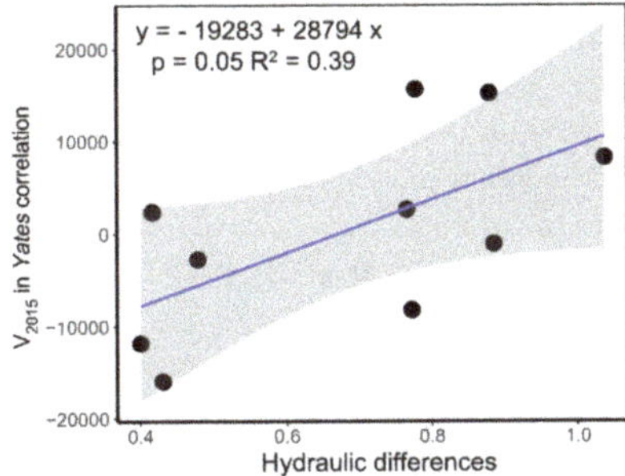

Figure 6. Relationship between hydraulic differences of each pair of species and their interspecific associations, represented by Canberra distance and V_{2015} in Yates correlation tests, respectively. Blue line shows the linear model result, while the gray region shows the confidence interval.

4. Discussion

Divergent hydraulic strategies can partly explain interspecific associations among tree species co-occurring in the temperate forest–steppe ecotone. The overall species associations within the forest were not significant in either years, suggesting the absence of species mutual exclusion caused by intensive interspecific competition in the plant community. The divergent hydraulic strategies illustrated that different water-use strategies may be a way trees avoid competition over water absorption, which was the primary limitation for the forests in forest–steppe ecotone [8,27]. The strong negative association within the forest occurred between *P. davidiana* and *B. platyphylla*, as well as *P. davidiana* and *B. dahurica*, indicating the strong interspecific competitions. The hydraulic strategies of these two species pairs were the least divergent ones. The efficient and safe hydraulic strategy of *T. mongolica* can explain the increase in abundance of *T. mongolica* during the studied years. In general, the positive relationship between interspecific association and hydraulic differences supported the previous hypotheses.

The divergent hydraulic strategies can reduce water stress among co-occurring species thus maintain the stability and species diversity of the forest community in drying forest–steppe ecotone. At the species level, divergent hydraulic strategies can reduce the interspecific competition for water resources. Species with higher water-transport efficiency can adopt more aggressive water absorbing strategies: trees can absorb and store as much water as possible after rainfall. Alternatively, species with strong water deficit tolerance can make use of small amounts of soil water in the long term. At the community level, the co-occurrence of species with divergent hydraulic strategies in forest–steppe ecotone can enlarge the moisture niche for tree species and increase the maintenance of semiarid forests. Thus, the divergent hydraulic strategies allow semiarid forests to maintain the same level of tree diversity with very large regional water supply differences. For example, the tree diversity in our study area with mean annual precipitation of 350 mm is quite similar to the ones with mean annual precipitation of 500–650 mm [19]. The maintaining of species diversity is increasingly important given that the forest–steppe ecotone is facing a drier and more variable climate in the future [28,29]. Divergent hydraulic strategies allow species to occupy moisture niches with many kinds of physical conditions such as slopes, aspects and soil textures [30]. However, we still lack the knowledge to determine water absorbing, storing and utilizing patterns for each species, which is urgently needed for forest dynamic studies and forest managements. Thus, enhanced long-term monitoring of plant physiological traits is needed in the forest–steppe ecotone in the future.

In contrast to the divergent hydraulic strategies, co-occurring species have convergent sensitivity to early stages of drought. The convergent P_{12} values of all the species except *Q. mongolica* indicate that the threshold for plant water uptake in response to early stages of drought [26] may be determined by the environment instead of the species. The convergent drought sensitivity can be explained by the hypothesis that plants tend to maximize their physiological functions at the edge of their environmental water limitation, which was put forward in a global meta-analysis for the hydraulic safety margin [31]. Only the "riskiest" approach does allow trees to maximize their carbon gains and biomass increase under normal water supply [31]. Moreover, the species with less safe hydraulic strategies must have some other strategies for drought tolerance or resilience improving, as all the species are facing the same level of drought and are similarly sensitive to the early stages of drought. The suckering regeneration strategy [21,32,33] seems to be a more important strategy for *P. davidiana* than the hydraulic strategies for its long-term survival in forest–steppe ecotone. The severe tree mortality during 2011 and 2012 can be explained by the neither safe nor highly efficient hydraulic strategy of *P. davidiana* in our results, while the rapid regeneration after drought events ensured the regrowth phase of *P. davidiana* instead of the succession to other species [21].

5. Conclusions

Our results revealed the linkage between plant hydraulic strategies and the interspecific associations of co-occurring trees in forest–steppe ecotone. Though missing some environmental data to fully investigate the relationships between the tree species and environment water resource supply, this study pointed out the important rule of plant physiology in the forest composition and dynamics, laying the foundation for further scientific exploration and data collection.

Supplementary Materials: The following are available online at http://www.mdpi.com/1999-4907/11/9/942/s1, Figure S1: The relationship between stem hydraulic transportation efficiency and safety (as measured by the water potential of stems at 88% loss of stem conductivity, P88) of all species, Figure S2: Spearman correlation between the interspecific association pattern and the hydraulic tradeoff differences among species.

Author Contributions: Conceptualization, H.L. and J.D.; methodology, J.D.; investigation, J.D., C.X., Y.Q., X.Z., B.L. and Y.W.; resources, M.Z.; data curation, J.D.; writing—original draft preparation, J.D.; writing—review and editing, J.D.; visualization, J.D.; supervision, H.L.; project administration, H.L.; funding acquisition, H.L. All authors have read and agreed to the published version of the manuscript.

Funding: This research was funded by National Natural Science Foundation of China, Grant Number 41530747, 41790422.

Acknowledgments: We are grateful to Guangyou Hao for his teaching in laboratory measurements and permission to use his laboratory facilities.

Conflicts of Interest: The authors declare no conflict of interest.

References

1. Rotenberg, E.; Yakir, D. Contribution of semi-arid forests to the climate system. *Science* **2010**, *327*, 451–454. [CrossRef]
2. Xu, C.; Liu, H.; Anenkhonov, O.A.; Korolyuk, A.Y.; Sandanov, D.V.; Balsanova, L.D.; Naidanov, B.B.; Wu, X. Long-term forest resilience to climate change indicated by mortality, regeneration, and growth in semiarid southern Siberia. *Glob. Chang. Biol.* **2017**, *23*, 2370–2382. [CrossRef] [PubMed]
3. Allen, C.D.; Macalady, A.K.; Chenchouni, H.; Bachelet, D.; McDowell, N.; Vennetier, M.; Kitzberger, T.; Rigling, A.; Breshears, D.D.; Hogg, E.H.; et al. A global overview of drought and heat-induced tree mortality reveals emerging climate change risks for forests. *For. Ecol. Manag.* **2010**, *259*, 660–684. [CrossRef]
4. Erdos, L. The edge of two worlds: A new review and synthesis on Eurasian forest-steppes. *Appl. Veg. Sci.* **2018**, *21*, 345–362. [CrossRef]
5. Chun-yu, Y.; Shao-fei, L.I.U.; Li-fei, Y.U. Interspecific associations of dominant tree species with restoration of a karst forest. *J. Zhejiang AF Univ.* **2010**, *27*, 44–50.

6. Liu, Z.; Zhu, Y.; Wang, J.; Ma, W.; Meng, J. Species association of the dominant tree species in an old-growth forest and implications for enrichment planting for the restoration of natural degraded forest in subtropical China. *Forests* **2019**, *10*, 957. [CrossRef]

7. Xie, T.; Tian-Zhen, J.U.; Shi, H.X.; Fan, Z.H.; Yang, G.K.; Zhang, S.Z. Interspecific association of rare and endangered Pinus bungeana community in Xiaolongshan of Gansu. *Chin. J. Ecol.* **2010**, *29*, 448–453.

8. Liu, H.; Park Williams, A.; Allen, C.D.; Guo, D.; Wu, X.; Anenkhonov, O.A.; Liang, E.; Sandanov, D.V.; Yin, Y.; Qi, Z.; et al. Rapid warming accelerates tree growth decline in semi-arid forests of Inner Asia. *Glob. Chang. Biol.* **2013**, *19*, 2500–2510. [CrossRef]

9. Nardini, A.; Pitt, F. Drought resistance of Quercus pubescens as a function of root hydraulic conductance, xylem embolism and hydraulic architecture. *New Phytol.* **1999**, *143*, 485–493. [CrossRef]

10. Zimmermann, M.H. Hydraulic architecture of some diffuse-porous trees. *Can. J. Bot.* **1978**, *56*, 2286–2295. [CrossRef]

11. Carvalho, E.C.D.; Martins, F.R.; Soares, A.A.; Oliveira, R.S.; Muniz, C.R.; Araujo, F.S. Hydraulic architecture of lianas in a semiarid climate: Efficiency or safety? *Acta Bot. Bras.* **2015**, *29*, 198–206. [CrossRef]

12. Sperry, J.S.; Meinzer, F.C.; Mcculloh, K.A. Safety and efficiency conflicts in hydraulic architecture: Scaling from tissues to trees. *Plant Cell Environ.* **2008**, *31*, 632–645. [CrossRef] [PubMed]

13. Kondoh, S.; Yahata, H.; Nakashizuka, T.; Kondoh, M. Interspecific variation in vessel size, growth and drought tolerance of broad-leaved trees in semi-arid regions of Kenya. *Tree Physiol.* **2006**, *26*, 899–904. [CrossRef] [PubMed]

14. Cruiziat, P.; Cochard, H.; Ameglio, T. Hydraulic architecture of trees: Main concepts and results. *Ann. For. Sci.* **2002**, *59*, 723–752. [CrossRef]

15. Hao, G.Y.; Lucero, M.E.; Sanderson, S.C.; Zacharias, E.H.; Holbrook, N.M. Polyploidy enhances the occupation of heterogeneous environments through hydraulic related trade-offs in Atriplex canescens (*Chenopodiaceae*). *New Phytol.* **2013**, *197*, 970–978. [CrossRef]

16. Choat, B.; Brodribb, T.J.; Brodersen, C.R.; Duursma, R.A.; Lopez, R.; Medlyn, B.E. Triggers of tree mortality under drought. *Nature* **2018**, *558*, 531–539. [CrossRef]

17. Sperry, J.S.; Love, D.M. What plant hydraulics can tell us about responses to climate-change droughts. *New Phytol.* **2015**, *207*, 14–27. [CrossRef]

18. Gebauer, R.L.E.; Schwinning, S.; Ehleringer, J.R. Interspecific competition and resource pulse utilization in a cold desert community. *Ecology* **2002**, *83*, 2602–2616. [CrossRef]

19. Jiang, Z.H.; Ma, K.M.; Anand, M.; Zhang, Y.X. Interplay of temperature and woody cover shapes herb communities along an elevational gradient in a temperate forest in Beijing, China. *Community Ecol.* **2015**, *16*, 215–222. [CrossRef]

20. Zeng, N.; Yao, H.X.; Zhou, M.; Zhao, P.W.; Dech, J.P.; Zhang, B.; Lu, X. Species-specific determinants of mortality and recruitment in the forest-steppe ecotone of northeast China. *For. Chron.* **2016**, *92*, 336–344. [CrossRef]

21. Zhao, P.W.; Xu, C.Y.; Zhoul, M.; Zhang, B.; Ge, P.; Zeng, N.; Liu, H.Y. Rapid regeneration offsets losses from warming-induced tree mortality in an aspen dominated broad-leaved forest in northern China. *PLoS ONE* **2018**, *13*, e0195630. [CrossRef] [PubMed]

22. Wheeler, J.K.; Huggett, B.A.; Tofte, A.N.; Rockwell, F.E.; Holbrook, N.M. Cutting xylem under tension or supersaturated with gas can generate PLC and the appearance of rapid recovery from embolism. *Plant Cell Environ.* **2013**, *36*, 1938–1949. [CrossRef] [PubMed]

23. Venturas, M.D.; MacKinnon, E.D.; Jacobsen, A.L.; Pratt, R.B. Excising stem samples underwater at native tension does not induce xylem cavitation. *Plant Cell Environ.* **38**, 1060–1068. [CrossRef] [PubMed]

24. Zwieniecki, M.A.; Holbrook, N.M. Diurnal variation in xylem hydraulic conductivity in white ash (*Fraxinus americana* L.), red maple (*Acer rubrum* L,) and red spruce (*Picea rubens Sarg,*). *Plant Cell Environ.* **1998**, *21*, 1173–1180. [CrossRef]

25. Alder, N.N.; Pockman, W.T.; Sperry, J.S.; Nuismer, S. Use of centrifugal force in the study of xylem cavitation. *J. Exp. Bot.* **1997**, *48*, 665–674. [CrossRef]

26. Bartlett, M.K.; Klein, T.; Jansen, S.; Choat, B.; Sack, L. The correlations and sequence of plant stomatal, hydraulic, and wilting responses to drought. *Proc. Natl. Acad. Sci. USA* **2016**, *113*, 13098–13103. [CrossRef]

27. Dulamsuren, C.; Hauck, M.; Leuschner, C. Recent drought stress leads to growth reductions in Larix sibirica in the western Khentey, Mongolia. *Glob. Chang. Biol.* **2010**, *16*, 3024–3035. [CrossRef]

28. Seager, R.; Vecchi, G.A. Greenhouse warming and the 21st century hydroclimate of southwestern North America. *Proc. Natl. Acad. Sci. USA* **2010**, *107*, 21277–21282. [CrossRef]
29. Piao, S.L.; Ciais, P.; Huang, Y.; Shen, Z.H.; Peng, S.S.; Li, J.S.; Zhou, L.P.; Liu, H.Y.; Ma, Y.C.; Ding, Y.H.; et al. The impacts of climate change on water resources and agriculture in China. *Nature* **2010**, *467*, 43–51. [CrossRef]
30. Chesson, P.; Gebauer, R.L.E.; Schwinning, S.; Huntly, N.; Wiegand, K.; Ernest, M.S.K.; Sher, A.; Novoplansky, A.; Weltzin, J.F. Resource pulses, species interactions, and diversity maintenance in arid and semi-arid environments. *Oecologia* **2004**, *141*, 236–253. [CrossRef]
31. Choat, B.; Jansen, S.; Brodribb, T.J.; Cochard, H.; Delzon, S.; Bhaskar, R.; Bucci, S.J.; Feild, T.S.; Gleason, S.M.; Hacke, U.G.; et al. Global convergence in the vulnerability of forests to drought. *Nature* **2012**, *491*, 752–755. [CrossRef] [PubMed]
32. Landhausser, S.M.; Wachowski, J.; Lieffers, V.J. Transfer of live aspen root fragments, an effective tool for large-scale boreal forest reclamation. *Can. J For. Res* **2015**, *45*, 1056–1064. [CrossRef]
33. Mitton, J.B.; Grant, M.C. Genetic variation and the natural history of quaking aspen. *Bioscience* **1996**, *46*, 25–31. [CrossRef]

 *forests*

Article

Do Sandy Grasslands along the Danube in the Carpathian Basin Preserve the Memory of Forest-Steppes?

Károly Penksza [1], Dénes Saláta [2], Gergely Pápay [1,*], Norbert Péter [3], Zoltán Bajor [2], Zsuzsa Lisztes-Szabó [4], Attila Fűrész [5], Márta Fuchs [6] and Erika Michéli [6]

[1] Institute of Crop Production Sciences, Szent István University, Páter Károly utca 1, H-2100 Gödöllő, Hungary; penksza@gmail.com

[2] Institute of Nature Resource Conservation, Szent István University, Páter Károly utca 1, H-2100 Gödöllő, Hungary; salata.denes@szie.hu (D.S.); Bajor.zoltan@szie.hu (Z.B.)

[3] Doctoral School of Environmental Sciences, Szent István University, Páter Károly utca 1, H-2100 Gödöllő, Hungary; peter.norbert87@gmail.com

[4] Institute for Nuclear Research, Isotope Climatology and Environmental Research Centre, Bem tér 18/c, H-4026 Debrecen, Hungary; lisztes-szabo-zsuzsanna@atomki.mta.hu

[5] Faculty of Agricultural and Environmental Sciences, Szent István University, Páter Károly utca 1, H-2100 Gödöllő, Hungary; furatis1@gmail.com

[6] Institute of Environmental Sciences, Szent István University, Páter Károly utca 1, H-2100 Gödöllő, Hungary; fuchs.marta@szie.hu (M.F.); micheli.erika@szie.hu (E.M.)

* Correspondence: geri.papay@gmail.com

Received: 16 November 2020; Accepted: 19 January 2021; Published: 20 January 2021

Abstract: Research highlights: In the present survey we examined the sandy grasslands appearing in the steppe-forest-steppe vegetation in the central part of the Carpathian Basin along the Danube. Background and objectives: We aimed to answer the following questions: Is it possible to build a picture of the past form of the vegetation through the examination of these vegetation units based on dominant grass taxa? Is *Festuca wagneri* an element of open grasslands or steppes? According to our hypothesis, these surveys can help reveal the original or secondary woody, shrubby patches through clarifying dominant taxa. Materials and Methods: We studied the grasslands in terms of coenology, putting great emphasis on the dominant *Festuca* taxa. Based on our preliminary surveys and literature, three vegetation types can be separated based on one single dominant *Festuca* taxon in each. The survey was conducted in four different locations in the Carpathian Basin. The cover of dominant grass species was used as an indicator value. The pedological background was also examined. Results: *F. vaginata* grassland is an open vegetation type based on its coenosystematic composition and ecological values. It grows in very weakly developed calcareous soil with sandy texture, with its lowest and highest organic carbon content ranging from 0.2% to 11.3% (0.2%), and the highest carbonate content (11.3%). Where the grasslands were disturbed, *F. pseudovaginata* and the recently discovered *F. tomanii* appeared. These taxa were also found in forest patches. The soil under *F. pseudovaginata* was more developed, in the surface horizon with higher organic carbon content (1.1%) and lower carbonate content (6.9%). The soil profile under *F. wagneri* developed the most, as the presence of deep and humus rich soil material from deflation and degradation showed. Conclusions: the dominant *Festuca* taxa of these vegetation types are good indicators of the changes in the vegetation and their ecological background.

Keywords: *Festuca vaginata*; *Festuca pseudovaginata*; *Festuca wagneri*; ecological values; pedological analysis; diversity

1. Introduction

We analyzed the sandy vegetation along the Danube in the central area of the Carpathian Basin, notably the calcareous sandy grasslands, which neighbor forest-steppe patches. The Pannonian–Pontic environmental zone (PAN) of the Carpathian basin, the Middle and Lower-Danube Plains and the Black-Sea. This area within characterized by natural forest-steppe and steppe vegetation [1–3]. The western borders of the Palaearctic steppe zone stretches across Eastern Europe, with high coverages of steppe and steppe-like grasslands in Bulgaria, Hungary, Moldova and Ukraine. The most important grassland types and subtypes in Eastern Europe [4] are the steppe grasslands (*Festuco-Brometea: Festucetalia valesiacae*), which are primary grasslands in the Eastern European region associated with the steppe and forest steppe zones. They are typically distributed in lowlands and at the foothills, and characterized by the dominance of *Festuca* and *Stipa* species.

The origin and history of the vegetation in the Pannonian zone are not always clear [5]. In the last centuries the original zonal appearance of the vegetation went through significant changes and became mosaic-like [6]. In the Carpathian Basin the zonal arrangement of the soil types and climate disbands completely, giving the place a mosaic-like landscape [6,7]. As a result, a mosaic-like vegetation types appeared. In areas with calcareous soil in the central area of the Carpathian Basin the environmental factors are also mosaic-like [8–10]. According to [11], in the sandy areas of the Danube-Tisza Interfluve, different vegetation patches are formed along environmental gradients. These gradients can be physical parameters, such as soil moisture, soil structure, exposure or temperature [12–14].

According to [15], in natural forest steppe transitions, trees can have a major effect on herb vegetation composition. In the understory of the forest and grove patches the species composition is different but instead of being a refuge of forest-specialized herbs other grassland-specialized species appear, contradicting the stress-gradient hypothesis [16,17], which states that competitive partners can change to facilitative ones at higher stress levels.

The regeneration potential of steppes is considerably lower than that of wet meadows, so mowing alone or intensive grazing is not effective in maintaining the diversity of the vegetation, though its more effective than abandoning it [18,19]; complex methods of low-level grazing and mowing should be used [20]. Ref. [21] also mention that richness of the soil can be a negative factor in the regeneration of arid steppes, since on more fertile soils the ruderal competitors make the natural recovery of specialist species nearly impossible.

According to [22,23], 45% of woody cover can be considered to be a threshold level, above which grassland specialist herbs give way to forest-related ones. They state that forest-steppe mosaics, closed forests and open grasslands also form a mosaic on higher altitudes, which also has to be taken into consideration when the management and conversation of these complex habitats are planned.

Refs. [24,25] summed up the main features and characteristics of Eurasian forest-steppes in an extensive review in which they describe the highly diverse fine-scale grassland-forest mosaics in the Carpathian basin as forest-steppes with characteristic species (among others) such as *Quercus* spp., *Acer tataricum*, *Populus alba* and herbs like *Chrispopgon gryllus*, *Festuca vaginata*, *F. rupicola*, *F. valesiaca*, *Stipa* spp. and *Astragalus* spp.

In the past few centuries, significant changes occurred in the relationship of forests and grasslands between the Danube and the Tisza. Grasslands appeared to replace forests and some swamps disappeared [26,27]. In the contact zones or rather, according to the present results, in places once covered in forests, open sandy surfaces formed then became inhabited by vegetation in a relatively short period of time [21,28–30]. At the same time, effects of human interventions can be detected long after they ended [31]. In places once covered in forests succession begins as a natural way of the regeneration of vegetation [32–36].

The soil-vegetation relationship in the sandy regions of the Danube–Tisza Interfluve was described and assessed by [37,38]. Ref. [39] also performed a botanical study of the Juniperetum of the Bugac area, in which he characterized the different vegetation types, compiled flora descriptions supplemented with coenological data and provided a guideline for the evaluation of the location of the vegetation types of the Hungarian Great Plain. Járó [40] (1974) summarized the types of habitats of the Danube–Tisza Interfluve and described soils as an important factor in determining its development.

The close surface geology of the Danube–Tisza Interfluve region is determined by mainly carbonatic eolian sediments of the Danube River. Based on the grain size distribution of the parent material (from coarse, medium and fine sand to the silty loess) and the hydraulic conditions (particularly the depth and quality of the water table) different soil types were developed in the area. At the high and dry landscape positions carbonatic shifting sands and humic sandy soils (Arenosols) can be found, which are characterized by unfavorable physical and chemical properties (high permeability and low water and nutrient storage capacity), thus they have low fertility [41,42].

Due to the regular redistribution of the eolian sandy deposits in the past buried soil horizons are quite frequent in the sandy soils of the region. The properties (i.e., grain size distribution, organic carbon content or the present soil structure) and the depth of the buried horizons may improve the fertility of the surface sandy soils and can provide information of the former soil forming environment as well [42]. The presence of soil structure or the different colors at different depths of the sands are also good indicators of former soil development under different conditions and vegetation cover of the past [43,44].

The climate of the region is continental with a sub-Mediterranean influence. The annual precipitation is 500–600 mm (maximum in June) with a mean temperature of approximately 11 °C, and increasing aridity from north to south [45–47].

Ref. [48] separated several new series within the *F. ovina* group. *F. vaginata* and *F. pseudovaginata* belong to the *F. psammophila* series [49,50]. The other newly described series is *F. trachyphylla*, which *F. tomanii* belongs to, according to its morphological features [51].

We posed the following questions: (i) How does the present vegetation reflect the original, natural vegetation? (ii) Are there any proof that there were forests patches in the area? (iii) Could the present grassland vegetation be a hint of the forest-steppe character? (iv) What inferences can be drawn when paralleling the present state of the vegetation with soil data?

In order to answer these questions we analyzed the pedological background of the vegetation types and used the survey of [52], which is the longest examination of sandy grasslands in the Pannonian Region, being conducted after shrub cutting and afforestation for 14 years continuously.

2. Materials and Methods

Coenological records were made in the central part of the Carpathian Basin, in 4 geographic units from northwest towards south and southeast. In the 4 areas dominant *Festuca* taxa were *Festuca vaginata, F. pseudovaginata* and *F. wagneri*, which were used as a baseline when differentiating records. The selected grasslands stretch along the Danube (Figure 1): 1. Little Hungarian Plain, Csallóköz; 2: northern part of the central area of the Carparthian Basin (Kiskunság); 3: southern part of the latter (Kiskunság) and 4: the southernmost sandy area of the Basin (Deliblát). Preferably, we chose sample areas on 3 different plain in each vegetation type.

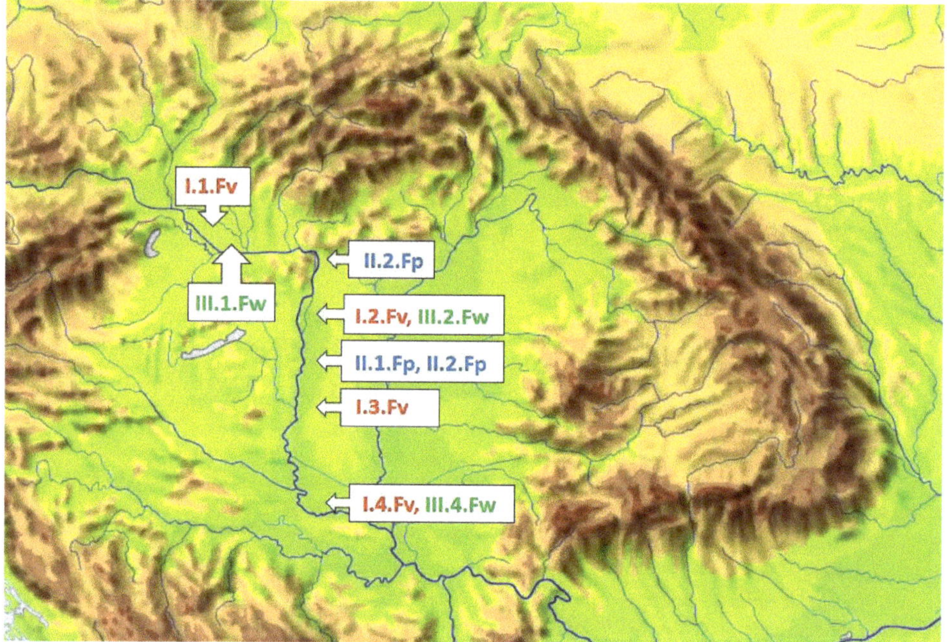

Figure 1. Location of the sample areas.

Taking this into account, our sample areas were the following:

Festuca vaginata grows everywhere along the Danube, and it appears in every studied geographic units. We could examine 3 sample areas in each northern part: the Little Hungarian Plain (I.1.Fv); northern part of Kiskunság (I.2.Fv) and southern part of Kiskunság (I.3.Fv). On the southernmost part (Deliblato, Serbia) only 1 sample area could be analyzed (I.4.Fv).

Festuca pseudovaginata grows only in the Carpathian Basin, on the northern plane. We examined 3 vegetation types dominated by it, based on the clearly visible physiognomical differences. The first one was a degraded type dominated by weeds at Vácrátót (II.1.Fp): The other one was more diverse, containing also arboreal species at Újpest (II.2.Fp): the third one was a natural grassland at Kunpeszér-Kunadacs (II.3.Fp).

Festuca wagneri was also found everywhere along the Danube in the Pannonian Region of the Carpathian Basin: in the Csallóköz (III.1.Fw), Northern part of Kiskunság (III.2.Fw), Southern part of Kiskunság (III.3.Fw) and in the southernmost part, at Deliblát (III.4.Fw).

The following relative ecological indicators [53] were used: relative temperature requirements (TB):

1: Subnaval or supraphoric belt;
2: Alpine, boreal or tundra belt;
3: Subalpine or subboreal belt;
4: Montana coniferous forest belt or taiga belt;
5: Belt of Montana deciduous mesophilic forests;
6: Belt of submontane deciduous forests;
7: Belt of thermophilic forests and forest steppes;

8: Sub-Mediterranean shiblets and steppe belt;

9: Plants of the Eumed Mediterranean evergreen zone.

Relative soil moisture requirements (WB):

1: Plants with high drought tolerance often in areas that are completely dehydrated or persistently extremely dry (rocky, semi-desert);

2: Drought indicating plants in long dry season production areas;

3: Drought-tolerant plants, occasionally found in fresh production areas;

4: Semi-arid crops;

5: Plants of semi-natural habitats;

6: Fresh crops;

7: Moisture indicator plants, focused on well ventilated, non-wet soil;

8: Humidity indicator, but also tolerant of short flooding;

9: Groundwater signaling plants on gravely saturated (air-poor) soils;

10: Aquatic plants of areas with variable water status, dehydrating for a shorter period;

11: Aquatic organisms floating, rooted or floating;

12: Submerged aquatic plants.

Nitrogen requirements (NB).

1: Plants of sterile, extremely nutrient-poor areas (e.g., peat moss);

2: Highly nutrient-poor crops;

3: Moderately oligotrophic plants;

4: Plants of submesotrophic habitats;

5: Mesotrophic plants;

6: Moderately nutrient-rich plants;

7: Plants of nutrient-rich areas;

8: N-labeled plants in fertilized soils;

9: Over-fertilized hypertrophic habitats (shepherds' farms), plants of ruined soils.

Taxon nomenclature was used according to [54]. Association nomenclature was used according to [5]. Values of digesting extreme climatic conditions (continentality, KB) were also used based on the 9-level scale of [55], which was based on [56]:

1. Eu-oceanic species, occurs occasionally in Central Europe (not in Hungary);
2. Oceanic species, occurs mainly in Western end Western Central Europe;
3. Oceanic/suboceanic species, occurs mainly in Central Europe;
4. Suboceanic species, occurs mainly in Central Europe and occasionally in Eastern Europe;
5. Transitory types, with a slight suboceanic and subcontinental feature;
6. Subcontinental species, occurs mainly in Central and Eastern Europe;
7. Continental-subcontinental species, occurs mainly in Eastern Europe;
8. Continental species, occurs occasionally in Central Europe;
9. Eu-continental species, occurs mainly in Siberia and Eastern Europe (not in Central Europe).

Spatial heterogeneity of soil cover was investigated by the use of a Dutch auger soil sampler and 2 soil profiles were opened and described in order to characterize the soil types of the study area. Morphological descriptions and classification of soil profiles were made on site according to international standards [57,58]. Based on our survey of soil and vegetation cover, 3 sampling sites were selected. Composite soil samples from the depth of 0–15 and 15–30 cm were collected for laboratory analysis from each selected sites and soil parameters that might be connected to vegetation were determined. Soil pH was measured in 1:2.5 soil–water suspension and in 1 M KCl, $CaCO_3$ was obtained by the Scheibler Calcimeter, salt concentration was determined by measuring the electrical conductivity of saturated paste [59] and soil organic carbon content (%) was determined by the wet chemical oxidation method given by [60]. The Walkley and Black method [61] utilizes a specified volume of acidic dichromate solution reacting with a known quantity of soil in order to oxidize the organic carbon. The oxidation step is then followed by titration of the excess dichromate solution with ferrous sulfate, then the organic carbon content is calculated using the difference between the total volume of dichromate added and the volume titrated after reaction.

For data analysis and presenting the results, the PAST [62,63] statistical software was used. For comparing the vegetation of the different localities, multivariate hierarchic cluster analysis (UPGMA—unweighted pair-group average [64]) was conducted using Euclidean mean distance. In the present study the diversity of vegetation is particularly important, therefore after collecting, contracting the data based on vegetation types, they were also analyzed using Rényi diversity profiles [65].

3. Results

3.1. Coenosystematic Results

Based on the coenosystematic results (the role of each species within the association), association group *Festucon vaginatae* differed the most from the others (based on mostly *Stipa borysthenica, Alkanna tinctoria, Centaurea arenaria, Dinathus serotinus* and *Koleria glauca* (Figure 2). In *F. vaginata* grasslands the proportion of these open sandy grassland species was particularly high and varied between 40 and 70% in every sample area. In *F. pseudovaginata* grasslands these taxa covered only 10–20%. In *F. wagneri* grasslands these ratios were similar to those found in the central region, although in the northern (III.1.Fw, Csallóköz) and southern (III.4.Fw, Deliblato) areas they dropped under 10%.

The proportion of *Festuco-Brometea* Br.-Bl. et R. Tx. ex Klika et Hadač 1944 was higher on the outer edges of the examined region (I.1.Fv, III.4.Fw). Based on the common taxa of *Festucetalia valesiacae* and *vaginatae* it was clear that *F. pseudovaginata* grasslands showed a transition, especially in the area described as shrub-forest mosaic (II.2.Fp). There were substantial differences among the elements of subcontinental arid grasslands (*Festucetalia valesiacae* Br.-Bl. and R. Tx. ex Br.-Bl. 1949) with regard to their relative proportions, too. These elements, as key taxa of sandy steppes, occurred primarily in the northernmost *F. vaginata* grasslands in the Little Hungarian Plain. In *F. pseudovaginata* grasslands their proportions were higher in the shrub-forest area (II.2.Fp). In *F. wagneri* grasslands, their cover values were found to be the highest in the northern part of Danube-Tisza Interfluve and at Deliblato (III.2.Fw and III.4.Fw). In the Deliblato area the following species were found in large numbers: *Adonis vernalis, Carex humilis* and *Jurinea mollis. F. tomanii* was rated in *Festucetalia valesiacae* based on its occurrences. Elements of *Festucetum rupicolae* were found only in two sample areas: in the forest parts of *F. pseudovaginata* grasslands and in the Northern Kiskunság patches of *F. wagneri*.

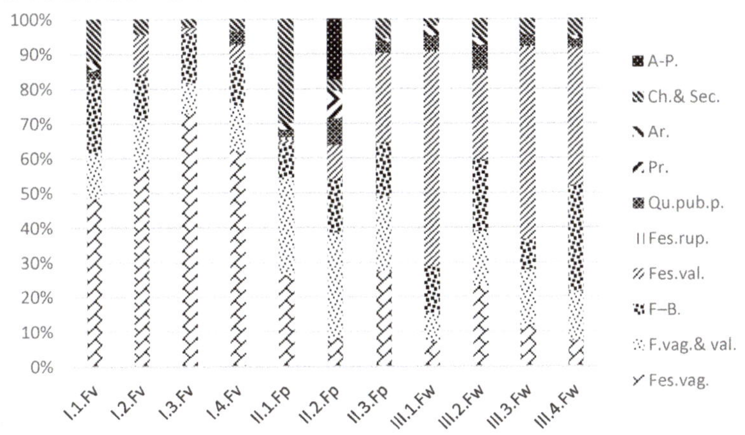

Figure 2. Coenosystematic composition of the grassland types under investigation. (*Festuca vaginata* vegetation types: I.1.Fv: Little Hungarian Plain, I.2.Fv: northern part of Kiskunság, I.3.Fv: southern part of Kiskunság, I.4.Fv southernmost part. *Festuca pseudovaginata* II.1.Fp: degraded type dominated by weeds, II.2.Fp: woody patches, II.3.Fp: natural patches; *Festuca wagneri* III.1.Fw: Csallóköz, III.2.Fw: Northern part of Kiskunság, III.3.Fw: Southern part of Kiskunság, III.4.Fw: in the southernmost part, Fes vag.: *Festucion vaginatae*, Fes. vag. and *Festucetalia vaginatae* and *valesiacae*, F-B.: *Festuco–Brometea*, Fes.val.: *Festucetalia valesiacae*, Fes.rup.: *Festucion rupicolae*, Qu.pub.p.: *Quercetea pubescentis-petreae*, P.: Prunion, Ar.: Arrhenatheretea, Ch. and Sec: Chenopodietea and Secaalietea, Onopordietea. A.-P.: Alno-Padion).

Weed vegetation elements (Chenopodietea and Secalietea Onopordietea) occurred mainly in weedy patches of *F. pseudovaginata* grasslands (II.1.Fp.) (i.e., *Anchusa officinalis, Aslepias syriaca, Carduus nutans, Portulaca oleracea, Setaria viridis* and *Tragus racemosus*). Taxa of European sub-Mediterranean and subcontinental forests occurred in the forest-steppe patches of *F. pseudovaginata* (*Quercetea pubescentis-petreae* (Oberd., 1948) Jakucs, 1960) (i.e., *Berberis viulgaris, Echinops schaerocephalon, Hierochloë repens* and *Veronica chamaedrys*). These elements were also found in *F. vaginata* grasslands, although with lower cover values, in the northernmost and southernmost areas. Forest taxa occur also in *F. wagneri* grasslands, mainly in the northern part of the Sand Ridge (III.2.Fw).

3.2. Diversity Results

Based on the Rényi diversity profile (Figure 3), it was clear that the most diverse vegetation type was *F. pseudovaginata*.

3.3. Ecological Values

We also analyzed the distribution of the relative ecological indices in the vegetation units, i.e., relative temperature, water requirements and continentality values.

F. vaginata grasslands were inhabited by species with the highest temperature requirement (value 8), although taxa with a value of 9 were to be found in secondary, weedy *F. pseudovaginata* grasslands (II.1.Fp). In *F. wagneri* grasslands, value 7 was the most common. On the two edges of the survey area, the coverages of half-shadow (value 5) and half-shadow–half-light (value 6) species were the highest. (Figure 4a).

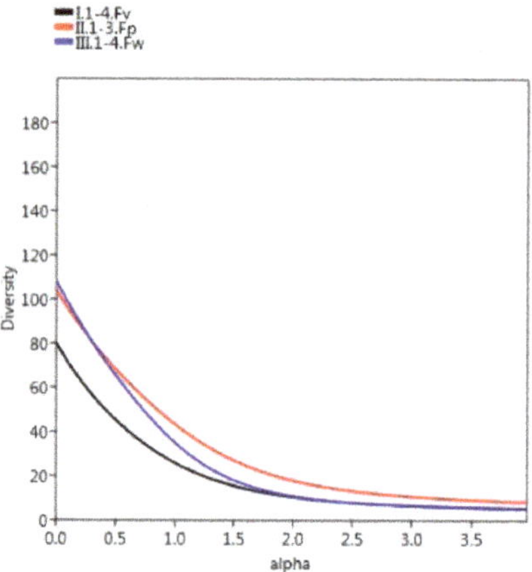

Figure 3. Rényi diversity of the grasslands (*Festuca vaginata* vegetation types: I.1-4.Fv, *Festuca pseudovaginata* II.1-3.Fp and *Festuca wagneri* III.1-4.Fw).

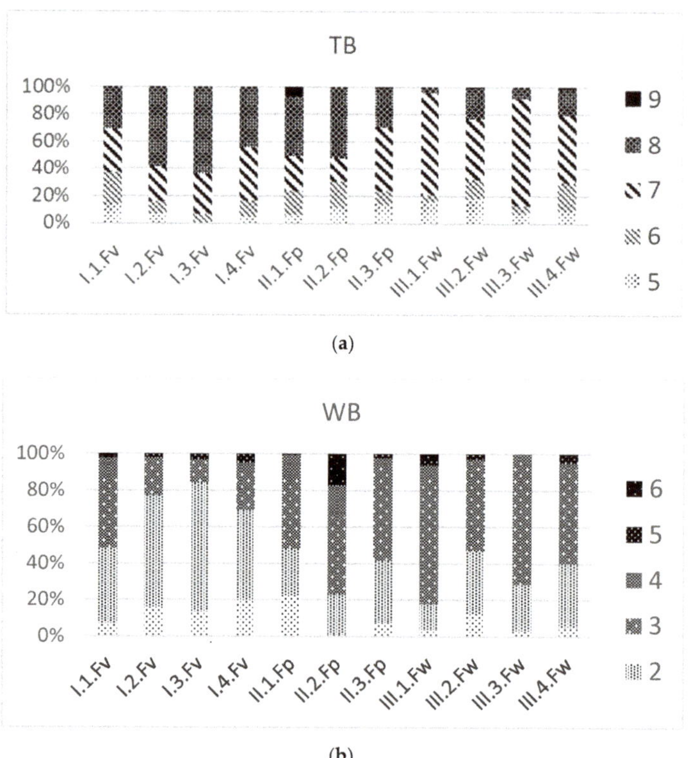

(**a**)

(**b**)

Figure 4. *Cont.*

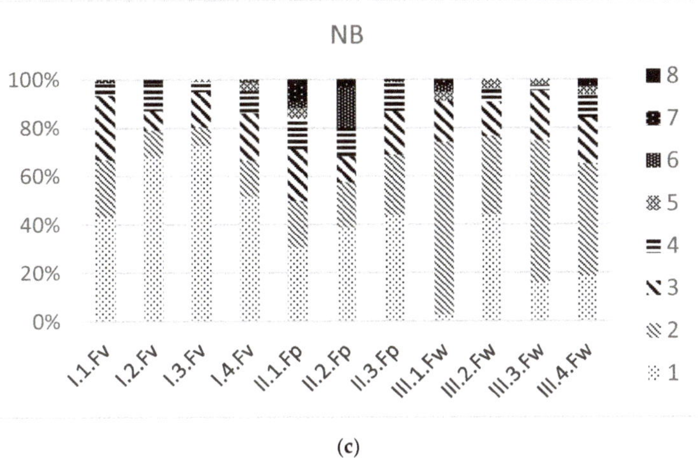

(c)

Figure 4. Proportions of the species based on relative ecological indices. (**a**) TB (relative temperature), (**b**) WB (relative water requirements) and (**c**) NB (nitrogen requirements). *Festuca vaginata* vegetation types: I.1.Fv: Little Hungarian Plain, I.2.Fv: northern part of Kiskunság, I.3.Fv: southern part of Kiskunság, I.4.Fv southernmost part. *Festuca pseudovaginata:* II.1.Fp: degraded type dominated by weeds, II.2.Fp: woody patches, II.3.Fp: natural patches. *Festuca wagneri:* III.1.Fw: Csallóköz, III.2.Fw: Northern part of Kiskunság, III.3.Fw: Southern part of Kiskunság, III.4.Fw: in the southernmost part. 1–9: see Materials and Methods.

Based on relative soil moisture requirements, *F. vaginata* differed from the others the most (Figure 4b). In the shrub-forest patches of *F. pseudovaginata* several species occurred, which indicated the borders of a wetter environment. Drought-tolerant species, which also occur in fresh habitats occasionally (value 3), were typical of *F. wagneri* grasslands.

Based on relative nitrogen requirements, *F. vaginata* differs again (Figure 4c). They contained the highest number of nutrient-poor patches. The shrub-forest patches of *F. pseudovaginata* were inhabited by species, which indicated larger amounts of nitrogen, which was also true for *F. wagneri* grasslands.

Species with continentality value of 9 (eucontinental), which occur very rarely in the Carpathian Basin, were to be found in the weedy grassland of *F. pseudovaginata* (II.1.Fp). Taxa with a value of 8, which are continental species marginally appearing in Central Europe, were also found. Value 7 (continental–subcontinental taxa with an Eastern European centre) had the largest proportion. Value 6 is a subcontinental category with a Central European centre; it also appeared along with value 5 (transitional types, with slight suboceanic and subcontinental features) (Figure 5).

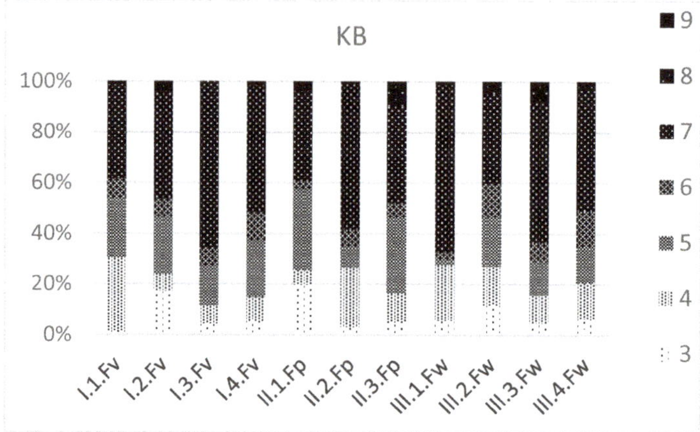

Figure 5. Proportion of the taxa based on their ability to tolerate extreme climatic effects. KB: continentality. *Festuca vaginata* vegetation types: I.1.Fv: Little Hungarian Plain, I.2.Fv: northern part of Kiskunság, I.3.Fv: southern part of Kiskunság, I.4.Fv southernmost part. *Festuca pseudovaginata* II.1.Fp: degraded type dominated by weeds, II.2.Fp: woody area, II.3.Fp: natural area. *Festuca wagneri* III.1.Fw: Csallóköz, III.2.Fw: Northern part of Kiskunság, III.3.Fw: Southern part of Kiskunság, III.4.Fw: in the southernmost part. 1–9: see Material and Methods.

3.4. Pedological Results

The soil conditions were quite homogenous in the studied areas but small scale differences in the rate of development, SOM (soil organic matter) and carbonate content were detected in the different *Festuca* spp habitats.

The soil type under *F. vaginata* was described as Calcaric Arenosol [59] with ACk Ck profile development [57], which represents a very weakly developed calcareous soil with sandy texture (Figure 6). Based on our results, *F. vaginata* usually occurs in areas with the least developed, strongly calcareous sandy soil patterns, where we measured the lowest organic carbon (0.2%), and the highest carbonate content (11.3%) (Table 1).

(**a**) (**b**) (**c**)

Figure 6. Soil profiles with genetic horizons [54] and classification according to the World Reference Base for Soil Resources [55]. (**a**) Calcaric Arenosol under *F. vaginata;* (**b**) Calcaric Arenosol under *F. pseudovaginata* and (**c**) Calcaric Brunic Arenosol under *F. wagneri.*

Table 1. Basic soil chemical parameters of the genetic horizons of the soils profiles sampled under *F. vaginata*, *F. pseudovaginata* and *F. wagneri* sites.

Site	Genetic Horizon	Depth (cm)	SOM (%)	CaCO$_3$ (%)	pH (H$_2$O)	pH (KCl)	Salt (%)
F. vaginata	ACk	0–5	0.2	11.3	8.2	8.0	0.02
	Ck	5-	0.1	11.4	8.6	8.2	0.02
F. pseudovaginata	Ak	0–5	1.1	6.9	8.1	7.7	0.03
	Bwk	5–18	0.6	8.6	8.3	8.1	0.03
	2Ck	18-	0.2	10.5	8.8	8.5	0.02
F. wagneri	A	0–20	1.3	0	6.7	6.4	0.01
	AB	20–60	0.8	0	6.9	6.1	0.01
	Bw	60–95	0.5	0	7.6	7.2	0.01
	BC	95–125	0.2	0	7.6	7.0	0.01
	2Ckl	125-	0.1	5.2	8.0	7.7	0.02

The soil under *F. pseudovaginata* was more developed with Ak Bwk 2Ck profile development [54] showing moderate accumulation of humified organic matter in the surface horizon with higher organic carbon content (1.1%) and lower carbonate content (6.9%) than in the topsoil of the *F. vaginata* habitat (Table 1).

The soil profile showed lithic discontinuity under the Ak Bwk horizons at 18 cm depth, indicated by the presence of common (5–10%) fine and medium (2–10 mm) rounded gravel content of the surface horizons compared to the underlying gravel free sand layer. The thin appearance of this surface layer can be explained by deforestation or other human impacts, which caused deflation and thus the loss of the former surface and surface close horizons. The recent, weakly developed and humus rich A horizon may have developed on the top of the remnant B horizon of the truncated soil, which was eroded to the surface.

As a result of this thin, weakly developed appearance of the described horizons, the soil under *F. pseudovaginata* was also classified as Calcaric Arenosol [55] but the morphology of the Bw horizon (at a depth of 5–18 cm) still supported our hypothesis of the presence of a former forest vegetation cover in the area. The olive brown (Munsell 2.5 YR 4/3 moist) color, and the weak granular structure are typical features of pedogenic alteration, which is characteristic for the weakly developed forest soils of the Carpathian Basin. These soils are called "Brown earths" in the genetic based Hungarian soil classification system [42], and "Cambisols" (or "Brunic Arenosols" in case of coarse sandy texture) according to the World Reference Base for Soil Resources [55].

The most developed soil profile of the studied area was described under *F. wagneri*. The solum was 125 cm thick with A AB Bw BC 2Ckl profile development and the soil was classified as Calcaric Brunic Arenosol [55] indicating obvious signs of soil formation with horizon differentiation in the sandy subsoil. Both the morphology and the chemical properties of the genetic horizons were found typical for soils developed under (former) forest vegetation. The soil showed relatively deep humus rich A horizon (0–20 cm), transitional character (AB horizon between 20–60 cm) and had a Bw horizon (at 60–95 cm depth) with brownish discoloration (Munsell 10 YR 4/4 moist), soil structure development (weak to moderate, medium subangular blocky) and leached, carbonate free characteristics. The presence of deep, transitional A and AB horizon indicated stabile surface with closed vegetation cover, which saved the humus rich soil material from deflation and degradation.

4. Discussion

Due to significant changes in the vegetation taking place in the last few hundred years, the central sandy grassland, forest-steppe areas of the Carpathian Basin became mosaic-like [6,66] but the present survey confirmed that several patches of the original vegetation was still preserved. At the same time, new opportunities opened up for new species and associations to settle, or even new taxa to form. For the same reason, invasive species could also invade more easily [31,67].

Based on the results we could confirm that *Festuca vaginata* was to be found all along the Danube in sandy soils in the Carpathian Basin but it is still uncertain whether its habitat spreads all the way to the Black Sea [5]. Based on our concurrent surveys it spreads only to Romania [6,68]. Basically, this species inhabits the open sandy grasslands, in the association of *Festucetum vaginatae* Rapaics ex Soó 1929 em. Borhidi 1996.

In earlier literature, *Festuca vaginata* was treated as the only dominant grass taxon of the open calcareous sandy grasslands. This was debated by [69] when finding *F. wagneri* in Hungary for the first time, although he treated it as a forest-steppe species. There was no consensus on the coenological affiliation of *F. wagneri*, although its taxonomy was clarified by [70] when describing it as a separate species. Since older specimens lose their epidermal hairs and their sclerenchime becomes annular, a greenish grass taxon was identified as F. wagneri in samples of *F. vaginata* grasslands, until [71] described it as a new species named *F. pseudovaginata*, which also forms an association new to science [72]. Later it was confirmed that the soil parameters of this association differ greatly from the others' [73] in terms of Ca and Mg contents. However, in the present survey soil profile was conducted for the first time, and its analysis confirmed what the environmental backdrop *F. pseudovaginata* indicated. This species evolved on forest soils. The soil profile showed 1.5 m deep forest soil and the amount of organic matter was higher. The relative ecological indices showed [50] that the vegetation type appeared under wetter conditions and based on their nitrogen requirement values, the soil was more nutrient-rich, which was also confirmed by the soil profiles.

F. pseudovaginata is endemic in the Pannonian region, it inhabits only the central sandy area of the Carpathian Basin. Ref. [74] provided data from the Romanian border, as its first occurrence in the country. However, there are still some uncertainties about it because *F. pseudovaginata* is tetraploid, while the specimens found were all diploid [75].

Coenosystematic analysis showed *F. pseudovaginata* mainly in forest-shrub areas and the samples also contained elements of *Quercetea pubescentis-petreae* and steppe taxa. In *F. wagneri* grasslands, proportion of taxa of *Festucetalia valesiacae* and *Festuco-Brometea* was higher. In addition, all three vegetation types were less diverse at their northern and southern edges and contained also forest, steppe and closed grassland species in larger proportions.

The present study was not extended to examine the border areas between sandy grasslands and forest vegetation, similarly to numerous invaluable and detailed works [23–25,76,77], although they did not examine the dominant *Festuca* taxa in detail. The present survey is part of a series of investigations, which examine the sandy grasslands along the Danube and the taxonomic conditions of the *Festuca*. As a result of our present study a new species was also discovered. Based on the soil profiles, it was clarified that these *Festuca* taxa can be used as indicators of disturbances in the vegetation or the onetime locations of forest patches.

F. pseudovaginata forms an association [72], not only under artificial circumstances as seen in Újpest but also at Kunpeszér-Kunadacs, where the vegetation indicates natural processes. The later is the largest continuous sandy grassland along the Danube in the Hungarian Plain, where the chances for vegetation types to evolve were the highest and its diverse environment gave an opportunity for mosaic-like landscape structure to form.

In conclusion, we answered our questions as follows: (i) According to our results, *F. vaginata* and *F. wagneri* grasslands can be considered as seminatural habitat based on their species composition and ecological indicators. *F. pseudovaginata* grasslands are mainly disturbed but natural patches were also found. (ii) Soils of *F. pseudovaginata* and *F. wagneri* indicate the onetime forest patches. (iii) The composition vegetation indicates its relationship with forest-steppes. Coenosystematic analysis showed *F. pseudovaginata* mainly in forest-shrub areas and the samples also contained elements of *Quercetea pubescentis-petreae* and steppe taxa. In *F. wagneri* grasslands, the proportions of taxa of *Festucetalia valesiacae* and *Festuco-Brometea* were higher. In addition, all three vegetation types were less diverse at their northern and southern edges and contained also forest, steppe and closed grassland species in higher proportions. (iv) The vegetation follows the aridification of the climate fast and changes rapidly into dry sandy grasslands form. The pedological results showed the memories of the changes of the soil. Answering the question in the title, the dominant *Festuca* species and their proportions and cover values will be the memory that indicates the vegetation types of the past in the area.

Author Contributions: Conceptualization, K.P., methodology, K.P.; software, D.S., M.F.; validation, K.P.; formal analysis, G.P.; investigation, K.P., D.S., G.P., N.P., Z.B., Z.L.-S., A.F., M.F. and E.M.; resources, K.P.; data curation, G.P.; writing—original draft preparation, K.P., M.F., G.P.; writing—review and editing, G.P.; visualization, G.P.; supervision, K.P.; project administration, G.P.; funding acquisition, K.P. All authors have read and agreed to the published version of the manuscript.

Funding: This research was funded by OTKA K-125423.

Data Availability Statement: The data presented in this study are available on request from the corresponding author. The data are not publicly available because the OTKA project is not completed yet.

Conflicts of Interest: The authors declare no conflict of interest. The funders had no role in the design of the study; in the collection, analyses, or interpretation of data; in the writing of the manuscript, or in the decision to publish the results.

References

1.	Török, P.; Janišová, M.; Kuzemko, A.; Rūsiņa, A.; Stevanović, Z.D. Grassland of the World: Management and Conservation Grasslands, their Threats and Management in Eastern Europe. In *Grasslands of the World: Diversity*

Management Conservation*; Squires, V.R., Dengler, J., Hua, L., Feng, H., Eds.; CRC Press: London, UK, 2018; pp. 64–88.

2. Török, P.; Wesche, K.; Ambarli, D.; Kamp, J.; Dengler, J. Step(pe) up! Raising the profile of the Palaearctic natural grasslands. *Biodivers. Conserv.* **2016**, *25*, 2187–2195. [CrossRef]

3. Wesche, K.; Ambarlı, D.; Kamp, J.; Török, P.; Treiber, J.; Dengler, J. The Palaearctic steppe biome: A new synthesis. *Biodivers. Conserv.* **2016**, *25*, 2197–2231. [CrossRef]

4. Mucina, L.; Bültmann, H.; Dierßen, K.; Theurillat, J.-P.; Raus, T.; Čarni, A.; Šumberová, K.; Willner, W.; Dengler, J.; Tichý, L. Vegetation of Europe: Hierarchical floristic classification system of plant, bryophyte, lichen, and algal communities. *Appl. Veg. Sci.* **2016**, *19* (Suppl. 1), 1–264. [CrossRef]

5. Borhidi, A.; Kevey, B.; Lendvai, G.; Seregélyes, T. *Plant Communities of Hungary*; Akadémiai Kiadó: Budapest, Hungary, 2012.

6. Molnár, Z.; Biró, M.; Bartha, S.; Fekete, G. Past trends, present state and future prospects of Hungarian forest-steppes. In *Eurasian Steppes. Ecological Problems and Livelihoods in a Changing World*; Werger, M.J.A., van Staalduinen, M.A., Eds.; Springer: Dordrecht, The Netherlands; Heidelberg, Germany; New York, NY, USA; London, UK, 2012; pp. 209–252.

7. Varga, Z. Die Waldteppen des pannonischen Raumes aus biogeographischer Sicht. *Düsseldorfer Geobot. Kolloqu.* **1989**, *6*, 35–50.

8. Tölgyesi, C.; Körmöczi, L. Structural changes of a Pannonian grassland plant community in relation to the decrease of water availability. *Acta Bot. Hung.* **2012**, *54*, 413–431. [CrossRef]

9. Tölgyesi, C.; Bátori, Z.; Erdős, L. Using statistical tests on relative ecological indicator values to compare vegetation units–Different approaches and weighting methods. *Ecol. Indic.* **2014**, *36*, 441–446. [CrossRef]

10. Marosi, S.; Somogyi, S. *Magyarország kistájainak katasztere I-II*; MTA Földrajztudományi Kutató Intézet: Budapest, Hungary, 1990; 1023p.

11. Tölgyesi, C.; Török, P.; Kun, R.; Csathó, A.I.; Bátori, Z.; Erdős, L.; Vadász, C. Recovery of species richness lags behind functional recovery in restored grasslands. *Land Degrad. Dev.* **2019**, *30*, 1083–1094. [CrossRef]

12. Bartha, S.; Campetella, G.; Ruprecht, E.; Kun, A.; Házi, J.; Horváth, A.; Virágh, K.; Molnár, Z. Will inter-annual variability in sand grassland communities increase with climate change? *Community Ecol.* **2008**, *9*, 13–21. [CrossRef]

13. Bartha, S.; Molnár, Z.; Fekete, G. Patch dynamics in sand grasslands: Connecting primary and secondary succession. In *The KISKUN LTER, Long-Term Ecological Research in the Kiskunság*; Kovács-Láng, E., Molnár, E., Kröel-Dulay, G., Barabás, S., Eds.; Institute of Ecology and Botany: Vácrátót, Hungary, 2008; Volume 37–40.

14. Bátori, Z.; Lengyel, A.; Maróti, K.; Körmöczi, L.; Tölgyesi, C.; Bíró, A.; Tóth, M.; Kincses, Z.; Cseh, V.; Erdős, L. Microclimate-vegetation relationships in natural habitat islands: Species preservation and conservation perspectives. *Időjárás Q. J. Hung. Meteorol. Serv.* **2014**, *118*, 257–281.

15. Tölgyesi, C.; Valkó, O.; Deák, B.; Kelemen, A.; Bragina, T.M.; Gallé, R.; Erdős, L.; Bátori, Z. Tree–herb co-existence and community assembly in natural forest-steppe transitions. *Plant Ecol. Divers.* **2018**, *11*, 465–477. [CrossRef]

16. Blaser, W.J.; Sitters, J.; Hart, S.P.; Edwards, P.J.; Venterink, H.O. Facilitative or competitive effects of woody plants on understorey vegetation depend on N-fixation, canopy shape and rainfall. *J. Ecol.* **2013**, *101*, 1598–1603. [CrossRef]

17. Bertness, M.; Callaway, R.M. Positive interactions in communities. *Trends Ecol. Evol.* **1994**, *9*, 191–193. [CrossRef]

18. Collins, S.L.; Knapp, A.K.; Briggs, J.M.; Blair, J.M.; Steinauer, E.M. Modulation of diversity by grazing and mowing in native tallgrass prairie. *Science* **1998**, *280*, 745–747. [CrossRef] [PubMed]

19. Güsewell, S.; Le Nédic, C. Effects of winter mowing on vegetation succession in a lakeshore fen. *Appl. Veg. Sci.* **2004**, *7*, 41–48. [CrossRef]

20. Csecserits, A.; Rédei, T. Secondary succession on sandy old fields in Hungary. *Appl. Veg. Sci.* **2001**, *4*, 63–74. [CrossRef]

21. Tölgyesi, C.; Bátori, Z.; Erdős, L.; Gallé, R.; Körmöczi, L. Plant diversity patterns of a Hungarian steppe-wetland mosaic in relation to grazing regime and land use history (Muster der Phytodiversität in ungarischen

Steppen-Feuchtwiesen Mosaiken in Abhängigkeit von der Beweidungsintensität und Landnutzungsgeschichte). *Tuexenia* **2015**, *35*, 399–416.

22. Erdős, L.; Tölgyesi, C.; Bátori, Z.; Semenishchenkov, A.; Yu, A.; Magnes, M. The influence of forest/grassland proportion on the species composition, diversity and natural values of an Eastern Austrian forest-steppe. *Russ. J. Ecol.* **2017**, *48*, 350–357. [CrossRef]

23. Erdős, L.; Török, P.; Szitár, K.; Bátori, Z.; Tölgyesi, C.; Kiss, P.J.; Bede-Fazekas, Á.; Kröel-Dulay, G. Beyond the forest-grassland dichotomy: The gradient-like organization of habitats in forest-steppes. *Front. Plant Sci.* **2020**, *11*, 236. [CrossRef]

24. Erdős, L.; Kröel-Dulay, G.; Bátori, Z.; Kovács, B.; Németh, C.; Kiss, P.J.; Tölgyesi, C. Habitat heterogeneity as a key to high conservation value in forest-grassland mosaics. *Biol. Conserv.* **2018**, *226*, 72–80. [CrossRef]

25. Erdős, L.; Tölgyesi, C.; Horzse, M.; Tolnay, D.; Hurton, Á.; Schulcz, N.; Körmöczi, L.; Lengyel, A.; Bátori, Z. Habitat complexity of the Pannonian forest-steppe zone and its nature conservation implications. *Ecol. Complex.* **2014**, *17*, 107–118. [CrossRef]

26. Biró, M.; Révész, A.; Molnár, Z.; Horváth, F.; Czúcz, B. Regional habitat pattern of the Danube-Tisza Interfluve in Hungary II. *Acta Biol. Hung.* **2008**, *50*, 19–60. [CrossRef]

27. Czúcz, B.; Révész, A.; Horváth, F.; Bíró, M. Loss of seminatural grasslands in the Hungarian forest steppe zone in the last fifteen years: Causes and fragmentation patterns. In Proceedings of the 13th Annual IALE (UK) Conference: Planning, People and Practice: The Landscape Ecology of Sustainable Landscapes, Northampton, UK, 12 September 2005; McCollinn, D., Jackson, J.I., Eds.; University of Northampton: Northampton, UK, 2005; pp. 73–80.

28. Csecserits, A.; Czúcz, B.; Halassy, M.; Kröel-Dulay, G.; Rédei, T.; Szabó, R.; Szitár, K.; Török, K. Regeneration of sandy old fields in the forest steppe region of Hungary. *Plant Biosyst.* **2011**, *145*, 715–729. [CrossRef]

29. Bartha, S.; Zimmermann, Z.; Horváth, A.; Szentes, S.; Sutyinszki, Z.; Szabó, G.; Házi, J.; Komoly, C.; Penksza, K. High resolution vegetation assessment with beta-diversity—A moving window approach. *Agric. Inform.* **2011**, *2*, 1–9. [CrossRef]

30. Albert, Á.-J.; Kelemen, A.; Valkó, O.; Miglécz, T.; Csecserits, A.; Rédei, T.; Deák, B.; Tóthmérész, B.; Török, P. Trait-based analysis of spontaneous grassland recovery in sandy old-fields. *Appl. Veg. Sci.* **2014**, *17*, 214–224. [CrossRef]

31. Pándi, I.; Penksza, K.; Botta-Dukát, Z.; Kröel-Dulay, G. People move but cultivated plants stay: Abandoned farmsteads support the persistence and spread of alien plants. *Biodivers. Conserv.* **2014**, *23*, 1289–1302. [CrossRef]

32. Török, P.; Kelemen, A.; Valkó, O.; Deák, B.; Lukács, B.; Tóthmérész, B. Lucerne-dominated fields recover native grass diversity without intensive management actions. *J. Appl. Ecol.* **2011**, *48*, 257–264. [CrossRef]

33. Török, P.; Deák, B.; Vida, E.; Valkó, O.; Lengyel, S.; Tóthmérész, B. Restoring grassland biodiversity: Sowing low-diversity seed mixtures can lead to rapid favourable changes. *Biol. Conserv.* **2010**, *143*, 806–812. [CrossRef]

34. Prach, K.; Pyšek, P.; Šmilauer, P. Prediction of vegetation succession in human-disturbed habitats using an expert system. *Restor. Ecol.* **1999**, *7*, 15–23. [CrossRef]

35. Prach, K.; Pyšek, P.; Bastl, M. Spontaneous vegetation succession in human-disturbed habitats: A pattern across seres. *Appl. Veg. Sci.* **2001**, *4*, 83–88. [CrossRef]

36. Prach, K.; Bartha, S.; Joyce, C.B.; Pyšek, P.; van Diggelen, R.; Wiegleb, G. The role of spontaneous vegetation succession in ecosystem restoration: A perspective. *Appl. Veg. Sci.* **2001**, *4*, 111–114. [CrossRef]

37. Szodfridt, I. Termőhelytípusok és vegetáció kapcsolata a Duna-Tisza közi homokháton. [Correlation of site conditions and vegetation in the Duna-Tisza köze-Sand plants as indicators for the sand afforestation]. *Abstr. Bot.* **1974**, *2*, 35–37.

38. Szodfridt, I. Talajvíz és vegetáció kapcsolata a Duna-Tisza köze homokterületén. [Correlation of the soil water table and the vegetation]. *Abstr. Bot.* **1974**, *2*, 39–42.

39. Szodfridt, I. A Duna-Tisza közi homokhátság növénytársulásainak fatermőképessége. [Timber growing capacity of the woodlands in Duna-Tisza köze]. *Erdészettudományi Közlemények* **1989**, *10*, 99–105.

40. Járó, Z. A Duna-Tisza-közi homokhát termőhely típusai. *Abstr. Bot.* **1974**, *2*, 31–34.

41. Dövényi, Z. *Magyarország Kistájainak Katasztere*; MTA Földrajztudományi Kutatóintézet: Budapest, Hungary, 2010; 876p.

42. Stefanovits, P. A tájak talajviszonyai. In *Talajtan*; Stefanovits, P., Filep, G., Füleki, G., Eds.; Mezőgazda Kiadó: Budapest, Hungary, 2010.

43. Barczi, A.; Joó, K. The role of Kurgans in the palaeopedological and palaeoecological reconstruction of the Hungarian Great Plain. *Z. Geomorphol.* **2009**, *53*, 131–137. [CrossRef]

44. Barczi, A.; Joó, K.; Pető, Á.; Bucsi, T. Survey of the buried palaeosol under Lyukas–mound. *Eurasian Soil Sci.* **2006**, *39*, 133–140. [CrossRef]

45. Fekete, G.; Molnár, Z.; Kun, A.; Botta-Dukát, Z. On the structure of the Pannonian forest steppe: Grasslands on sand. *Acta Zool. Acad. Sci. Hung.* **2002**, *48*, 137–150.

46. Kun, A. Analysis of precipitation year types and their regional frequency distributions in the Danube-Tisza mid-region, Hungary. *Acta Bot. Hung.* **2002**, *43*, 175–187. [CrossRef]

47. Tölgyesi, C.; Zalatnai, M.; Erdős, L.; Bátori, Z.; Hupp, N.R.; Körmöczi, L. Unexpected ecotone dynamics of a sand dune vegetation complex following water table decline. *J. Plant Ecol.* **2016**, *9*, 40–50. [CrossRef]

48. Pawlus, M. Systematyka i rozmieszczenie gatunków grupy *Festuca ovina* L. w. Polsce [Classification and distribution of *Festuca ovina* L. in Poland]. *Fragm. Flor. Geobot.* **2005**, *29*, 219–295.

49. Penksza, K. Kiegészítések a hazai Festuca taxonok ismeretéhez I. A Festuca psammophila series Festuca vaginata alakkörei [Additions to the knowledge of Hungarian Festuca taxa I. Taxa of Festuca psammophila series in the Festuca vaginata species complex]. *Bot. Közlemények* **2019**, *106*, 65–70. [CrossRef]

50. Penksza, K.; Szabó, G.; Zimmermann, Z.; Lisztes-Szabó, Z.; Pápay, G.; Járdi, I.; Fűrész, A.; Falusi, S.-E. A Festuca vaginata alakkör taxonómiai problematikája és ennek cönoszisztematikai vonatkozásai [The taxonomic problems of the Festuca vaginata agg. and their coenosystematic aspects]. *Georg. Agric.* **2019**, *23*, 63–76.

51. Korneck, D.; Gregor, T. *Festuca tomanii* sp. nov., ein Dünen-Schwingel des nördlichen oberrhein-, des mittleren main- und des böhmischen Elbetales. *Kochia* **2015**, *9*, 37–58.

52. Bajor, Z.; Zimmermann, Z.; Szabó, G.; Fehér, Z.; Járdi, I.; Lampert, R.; Kerényi-Nagy, V.; Penksza, P.L.; Szabó, Z.; Székely, Z.; et al. Effect of conservation management practices on sand grassland vegetation in Budapest, Hungary. *Appl. Ecol. Environ. Res.* **2016**, *14*, 233–247. [CrossRef]

53. Borhidi, A. Social behaviour types, the naturalness and relative ecological indicator values of the higher plants in the Hungarian flora. *Acta Bot. Acad. Sci. Hung.* **1995**, *39*, 97–181.

54. Király, G. *Új Magyar Füvészkönyv. Magyarország Hajtásos Növényei (New Hungarian Herbal. The Vascular Plants of Hungary. Identification Key)*; Aggteleki Nemzeti Park: Jósvafő, Hungary, 2009. (In Hungarian)

55. Ellenberg, H. Zeigerwerte der Gefässpflanzen Mitteleuropas. *Scr. Geobot.* **1974**, *9*, 1–97.

56. Meusel, H.; Schubert, R. *Volk. und Wissen*; Akademie Verlag: Berlin, Germany, 1972.

57. FAO. *Guidelines for Soil Description*, 4th ed.; FAO: Rome, Italy, 2006.

58. IUSS Working Group WRB. *World Reference Base for Soil Resources 2014 (Updated 2015)*; World Soil Resources Reports 106; Food and Agriculture Organization (FAO) of United Nations: Rome, Italy, 2016.

59. Buzás, I. *Soil and Agrochemistry Analysing Method Book No. 1. Physical, Hydrological and Mineralogy Analyses of Soils*; INDA 4231 Publishing: Budapest, Hungary, 2013; p. 357. (In Hungarian)

60. Walkley, A.; Black, I.A. An examination of the Degtjareff method for determining soil organic matter and a proposed modification of the chromic acid titration method. *Soil Sci.* **1934**, *37*, 29–38. [CrossRef]

61. Nelson, D.W.; Sommers, L.E. Total carbon, organic carbon, and organic matter. Methods of Soil Analysis. Part 3. Chemical Methods. *Soil Sci. Soc. Am. Book Ser.* **1996**, *5*, 961–1010.

62. Hammer, Ø. *PAST–PAleontological STatictics Version 3.06 Reference Manual*; Natural History Museum, University of Oslo: Oslo, Norway, 2015; 225p.

63. Hammer, Ø.; Harper, D.A.T.; Ryan, P.D. PAST–Paleontological Statistics Software Package for Education and Data Analysis. *Palaeontol. Electron.* **2001**, *4*, 1–9.

64. Saitou, N.; Nei, M. The neighbor-joining method: A new method for reconstructing phylogenetic trees. *Mol. Biol. Evol.* **1994**, *4*, 406–425.

65. Tóthmérész, B. Comparison of different methods for diversity ordering. *J. Veg. Sci.* **1995**, *6*, 283–290. [CrossRef]

66. Bátori, Z.; Erdős, L.; Kelemen, A.; Deák, B.; Valkó, O.; Gallé, R.; Bragina, T.M.; Kiss, P.J.; Kröel-Dulay, G.; Tölgyesi, C. Diversity patterns in sandy forest-steppes: A comparative study from the western and central Palearctic. *Biodivers Conserv.* **2018**, *27*, 1011–1030. [CrossRef]

67. Csontos, P.; Bózsing, E.; Cseresnyés, I.; Penksza, K. Reproductive potential of Asclepias syriaca stands in the rural surroundings of Budapest, Hungary. *Pol. J. Ecol.* **2009**, *57*, 383–388.

68. Penksza, K.; Péter, N.; Saláta, D.; Pápay, G.; Lisztes-Szabó, Z.; Bajor, Z. Result of conservation management and restauration in open sandy grasslands in the Homoktövis Nature Conversation area (Budapest, Hungary). In *Proceedings of the International Conference on Veterinary, Agriculture and Life Sciences (ICVALS), Antalya, Turkey, 29 October–1 November 2020, Abstract Book*; Mehmet, O., Ed.; ISRES Publishing: Antalya, Turkey, 2020; p. 7.

69. Pócs, T. A rákoskereszturi "Akadémiai erdö" vegetációja. [Die Vegetation des "Akademischen Waldes" in Rákoskeresztur]. *Bot. Közlemények* **1954**, *45*, 283–294.

70. Penksza, K.; Engloner, A. Taxonomic study of Festuca wagneri (Degen Thaisz et Flatt) in Degen Thaisz et Flatt. 1905. *Acta Bot. Sci. Hung.* **2000**, *42*, 257–264.

71. Penksza, K. *Festuca pseudovaginata*, a new species from sandy areas of the carpathian basin. *Acta Bot. Hung.* **2003**, *45*, 365–372. [CrossRef]

72. Penksza, K.; Csontos, P.; Pápay, G. Festucetum pseudovaginatae ass. nova is a new indigenous association in the Pannonian basin. *Haquetia* **2020**, in press.

73. Szabó, G.; Zimmermann, Z.; Catorci, A.; Csontos, P.; Wichmann, B.; Szentes, S.; Barczi, A.; Penksza, K. Comparative study on grasslands dominated by Festuca vaginata and F. pseudovaginata in the Carpathian Basin. *Tuexenia* **2017**, *37*, 415–429.

74. Šmarda, P.; Šmerda, J.; Knoll, A.; Bureš, P.; Danihelka, J. Revision of Central European taxa of Festuca ser. Psammophilae Pawlus: Morphometrical, karyological and AFLP analysis. *Plant Syst. Evol.* **2007**, *266*, 197–232.

75. Fűrész, A.; Hurta, A.; Kiss, E.; Pápay, G.; Kovács, L.; Péter, N.; Lantos, C.; Penksza, K. Morphotaxonomic and ploidy analysis of dominant Festuca species in sandy grasslands along the Danube. In *62. Georgikon Napok Nemzetközi Tudományos Konferencia: A klímaváltozás kihívásai a következő évtizedekben: Előadások kivonatai*; Szent István Egyetem Georgikon Kar: Keszthely, Hungary, 2020; p. 35.

76. Erdős, L.; Krstonošić, D.; Kiss, P.J.; Bátori, Z.; Tölgyesi, C.; Škvorc, Ž. Plant composition and diversity at edges in a semi-natural forest–grassland mosaic. *Plant Ecol.* **2019**, *220*, 279–292. [CrossRef]

77. Erdős, L.; Tölgyesi, C.; Cseh, V.; Tolnay, D.; Cserhalmi, D.; Körmöczi, L.; Gellény, K.; Bátori, Z. Vegetation history, recent dynamics and future prospects of a Hungarian sandy forest-steppe reserve: Forest-grassland relations, tree species composition and size-class distribution. *Community Ecol.* **2015**, *16*, 95–105. [CrossRef]

Article

Habitat Mosaics of Sand Steppes and Forest-Steppes in the Ipoly Valley in Hungary

Ildikó Járdi [1,*], Dénes Saláta [1], Eszter S.-Falusi [1], Ferenc Stilling [1], Gergely Pápay [1], Zalán Zachar [1], Dominika Falvai [1], Péter Csontos [2], Norbert Péter [1] and Károly Penksza [1]

[1] Faculty of Agricultural and Environmental Sciences, Szent István University, Páter Károly u. 1., H-2100 Gödöllő, Hungary; Salata.Denes@szie.hu (D.S.); falueci@gmail.com (E.S.-F.); stillingf@gmail.com (F.S.); geri.papay@gmail.com (G.P.); kereskenyi1@gmail.com (Z.Z.); domi.falvai@gmail.com (D.F.); peter.norbert87@gmail.com (N.P.); penksza@gmail.com (K.P.)

[2] Centre for Agricultural Research, Institute for Soil Science and Agricultural Chemistry, Herman Ottó út 15, H-1022 Budapest, Hungary; cspeter@mail.iif.hu

* Correspondence: ildikojardi@gmail.com

Academic Editor: László Erdős

Received: 24 November 2020; Accepted: 20 January 2021; Published: 25 January 2021

Abstract: The present study focuses on the mosaic-like occurrences of patches of steppes and fore-steppes in the Pannonian forest-steppe zone. We present the current vegetation, which is maintained including by human landscape use, i.e., grazing and mowing. The area is complex and for this reason it shows the changes in the landscape and differences in the vegetation more diversely. We wanted to answer the questions: Do sand steppes and forest-steppes occur in the Ipoly Valley and what location? What kind of environmental effects influence the species composition on these areas? Besides classic habitat mapping, are the satellite data from Sentinel-2A useful for distinction of different areas? Comparison of vegetation patches was based on the Hungarian habitat classification system (ÁNÉR). Based on satellite images, quantile data of the Normalized Vegetation Index (NDVI) were used for comparison. Based on the result, water bodies and urban areas are clearly distinguishable from other natural habitats. In some natural vegetation types, we found visible differences, such as grasslands, i.e., sandy steppe meadows and shrubby, woody vegetation patches. Sandy vegetation mainly grows on calcareous soils, which appear to be mosaic-like in the landscape on raised alluvials on the patches of past islands and reefs. From open to continuous closed grasslands, these vegetation types mainly grow on lithosoils. New occurrences of Pannonian sandy vegetation were discovered. In the sandy areas along the Ipoly Valley, open sandy grasslands were found, which is where the northernmost known occurrences of this vegetation type are. Besides common sandy grassland species, the vegetation also contains herbs that are typical in loess-grasslands and it is maintained by grazing, similarly to the eastern Pannonian area. This type of grazing can be useful when maintaining the mosaic-like appearance and diversity of the vegetation.

Keywords: forest-steppes; sandy grassland; grazing-mowing; NDVI; Sentinel-2A

1. Introduction

The Pannonian-Pontic environmental zone (PAN) occupies the major part of the Carpathian Basin. The area is characterized by natural forest-steppe and steppe vegetation [1,2].

Numerous studies were carried out investigating the sandy areas of the Pannon region. The first significant review was published by Zólyomi [3]. Szujkóné–Lacza [4] assembled and reviewed the literature

Forests **2021**, *12*, 135; doi:10.3390/f12020135

and collections of the Danube-Tisza Interfluve, while also using the data of the Botanical Department of the Hungarian Natural History Museum. However, surveys usually dealt with the central region of Hungary, which includes the most natural (i.e., relatively intact) parts. Owing to these studies, the temporal structure, the distinguishable aspects, and aspect-forming species of *Festucetum vaginatae* Simon 2000 are well known [5].

The zonal arrangement of the soil types and climate, which are characteristic of the eastern part of the continent, disbands completely and gives way to a mosaic-like landscape in the Carpathian Basin [6–8].

Both climazonal and edaphic mosaic habitats from steppe patches to forested areas [9–11], developed in sandy soil, which was altered and restricted as a result of landscape managements [12,13].

In these areas with calcareous soil in the central parts of the Carpathian Basin, the environmental factors are mosaic-like as well [14–19]. The present study examines the occurrences of the steppe and forest-steppe vegetation on the edge of the Pannonian forest-steppe region. The river Ipoly is one of the last rivers that have been preserved in their natural condition and have not been affected by water flow regulations. Not surprisingly, the Ipoly Valley is a protected nature reserve area of national significance, since it is part of the Danube-Ipoly National Park. In addition, it is also a nature conservation area (HUDI20026) and bird sanctuary (HUDI10008) and is subject to the Ramsar Convention in order to protect the migratory aquatic birds [20,21].

Despite its linear nature, Ipoly Valley harbors especially mosaic-like vegetation, mainly because it is a non-regulated, natural watercourse [22]. Close relations between the soil moisture level and degree of vegetation heterogeneity were also detected in other watercourses in the Pannonian region [23]. Based on earlier examinations of the Ipoly Valley, the changes of the ground water table clearly affect the spatial arrangement of vegetation types [24]. With the increased advance of agriculture, most of the grasslands were drained and plowed. The area along the Ipoly offers an ideal research terrain for studying the effects of various environmental factors such as soil on the distribution of native species. For these reasons, further examination of this area is needed.

Járdi et al. compared the coenology of the acidic sandy grasslands, steppes, meadows, and swamp meadows, which were grazed by Charolais and Hungarian gray cattle. The species pool and the cover of common species differed greatly in the examined grasslands, which clearly showed effects of the different abiotic and pedologic factors, water supply, and landscape uses. In these sandy grasslands, *Festuca ovina* L. aggregate (Poaceae) and *Festuca rupicola* Heuff. both appeared, and both *Festuca pseudovina* Hack. ex Wiesb. and *Stipa borysthenica* Klokov ex Prokudin were common in the more arid areas [25].

Examinations can be expanded with data from satellite images [26]. Sentinel-2A is the most useful satellite for this goal [27]. It was launched on June 23, 2015 as part of the European Copernicus Program.

Our questions were the following:

(i)   Do sand steppes and forest-steppes occur in the Ipoly Valley and if yes, then where?
(ii) What kind of environmental effects influence the species composition on these areas?
(iii) Besides classic habitat mapping, are the satellite data from Sentinel-2A useful for distinction of different areas?

2. Materials and Methods

The study area covers 445 hectares and is situated in northern Hungary along the River Ipoly, in the municipality area of Dejtár, and a smaller proportion between Dejtár and Ipolyvece (Figure 1).

Manual GPS was used for recording coordinates of the sample points with consideration to the edges of the habitat patches, which were treated as separate units. For the identification of habitats, the protocol of the Hungarian Habitat Classification System (ÁNÉR) was used. This system was developed in connection with the National Biodiversity Monitoring System (NBmR), which contains all vegetation types in Hungary [29]. It is the most frequently used complex system in the country and is under continuous development. Compared to coenological systems, NBmR is substantially simpler, as it contains fewer and broader categories. It groups associations into larger, more interpretable types and it is also feasible for practical use in nature conservation [30,31]. Association names were used according to [32]. Species nomenclature was used according to [33].

In general, one demarcated habitat patch belongs to only one habitat type (e.g., P2b see below in Table 1). However, there were habitat patches in which more than one habitat type was present. The main reason for this was that in some cases, habitat patches could not be allocated from the habitats in presentable size or they appeared as a mosaic of two or more ÁNÉR categories, therefore these habitat patches were indicated as habitat complexes in decreasing order of share in the following manner: D34 × B1a × P2a. The relevant ÁNÉR categories are shown in Table 1. While editing the habitat map, it was important to clearly trace and evaluate changes of habitat types.

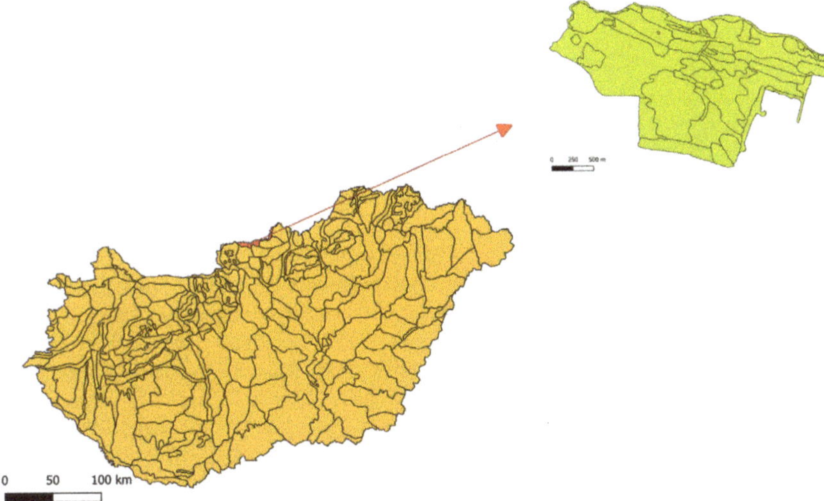

Figure 1. The location of the study area in Hungary (prepared with Marosi and Somogyi 1990) [28].

Table 1. Habitats revealed in the study area, using the categories of the Hungarian Habitat Classification System (ÁNÉR).

ÁNÉR Categories	Description
B1a	Eu- and mesotrophic reed and Typha beds
B5	Non-tussock tall-sedge beds
D34	Mesotrophic wet meadows
H5b	Closed sand steppes
P2b	Dry and semi-dry pioneer scrub
P2a	Wet and mesic pioneer scrub
J3	Riverine willow scrub
J4	Riverine willow-poplar woodlands
OB	Uncharacteristic mesic grasslands
OC	Uncharacteristic dry and semi-dry grasslands
P2b	Dry and semi-dry pioneer scrub
RB	Uncharacteristic or pioneer softwood forests
S2	Populus × euramericana plantations
S4	Scots and black pine plantations
T1	Annual intensive arable fields
U8	Water streams
U7	Sand, gravel, clay and peat mines, loess walls
U9	Standing waters
U11	Roads and railroads

Maps were created using QuantumGIS, an open source geoinformatical program, which can be downloaded from (www.qgis.org). Coordinates recorded in the field were visualized as a GSX file in the program. The Satellite images chosen for evaluation were recorded on 17th September 2019 as the cloud cover was relatively low and this date was as close to the field research's date as possible. All Sentinel-2A image was downloaded from the official homepage of Copernicus. From the downloaded 12 optical bands, two optical bands were used to extract the Normalized Vegetation Index (NDVI) data. The visible red (RED) and near-infrared (NIR) were used to compute the numerical Normalized Vegetation Indices (NDVI) of each 10×10 m pixel [26], then the pixels were colored accordingly. NDVI is a non-dimensional value that reflects the vegetational activity of a given area. It is returned by the quotient of the sum and the difference of the reflected intensity of NIR and RED [34]. NDVI shows the biological activity of the vegetation: the higher the reflection of the chlorophyll, the higher the value. In the absence of vegetation, NDVI will be negative, for example on water bodies in the early vegetation period. In the evaluation phase habitats mapped by using the classic field survey method were compared with NDVI of satellite images.

In order to visualize the latter, 20 points were selected randomly to each ÁNÉR category, then the NDVI of the points were assigned to the categories. Data were analyzed using Microsoft Excel and PAST (PAleontological STatistics) software [35].

Mean NDVI data were compared by the non-parametric Kruskal-Wallis test, since raw data of 5 habitat types out of the 19 habitat types involved in this analysis did not fit the Gaussian distribution. Dunn's Multiple Comparisons test was used as a post hoc test, and differences at level $p < 0.05$ were considered significant [35].

3. Results

3.1. Habitat Map with ÁNÉR Categories

Based on the field works, 19 habitat types (i.e., ÁNÉR categories) and their combinations (habitat complexes) were identified, resulting in a total of 29 patch types in the map. Table 2 summarizes the size and the number of the identified habitat patches.

Table 2. The number and the size of the habitat patches identified in the study area.

Type of Habitats	Number of Habitat Patches (Pieces)	Area (Hectares)
B1a	4	46.0
B2 × B5	1	17.5
D34	1	34.1
D34 × B1a × P2a	1	16.5
H5b	16	38.0
H5b × P2b	2	28.1
J3	1	6.8
J3 × B5	1	1.0
J3 × P2b × U9	1	0.9
J4	2	0.5
J4 × B5	1	3.3
J4 × P2b	7	25.7
J4 × P2b × U9	1	2.7
J4 × U8	2	11.3
OB	1	8.1
OC	8	30.6
P2b	15	18.0
P2b × D34	1	1.1
P2b × H5b	1	13.1
P2b × OC	1	0.2
RB	2	84.9
S2	5	30.4
S4	2	3.5
S7	2	6.2
T1	2	7.1
U11	3	6.8
U6	1	0.4
U7	1	0.7
U9	2	1.7
Total	88	445.1

3.2. Habitat Types of the Study Area

According to the habitat map (Figure 2), mosaics of gallery forests (J4) and swamp meadows (B1a, B2, B5) are common in the north-eastern edge of the area. Wet meadows are also common due to the water supply. Figure 3 shows the vegetation belt in the foreground of the gallery forest. It is seemingly continuous and its parts are difficult to distinguish during habitat mapping. From the narrow belt of the gallery forest (Figure 3A1) the following plant associations appear: *Phragmitetum vulgaris* Soó 1927, *Scirpetum lacustris* Chouard 1924, *Typhetum angustifoliae* Pignatti 1953, *Typhetum latifoliae* Lang 1973, *Sparganietum erecti* Roll 1938. In the wetter part (Figure 3A2), swamp meadows and *Carex*-dominanted associations appear as a complex, with patches of different sizes: *Caricetum gracilis Almquist* 1929, *Caricetum vesicariae* Chouard 1924, *Galio palustris-Caricetum ripariae* Bal.-Tul.et al.1993, *Caricetum vulpine* Soó 1927, *Caricetum melanostachyae*

Baláž 1943, *Caricetum distichae* Jonas 1933, and in some places even the *Phalaridetum arundinaceae* Libbert 1931, *Oenantho aquaticae-Rorippetum amphibiae* R.Tx.1953 and the *Butometum umbellati* Philippi 1973. In this belt, *Lychnis flos-cuculi* L. (Figure 3B) and *Ranunculus* ssp. are also dominant in the spring aspect.

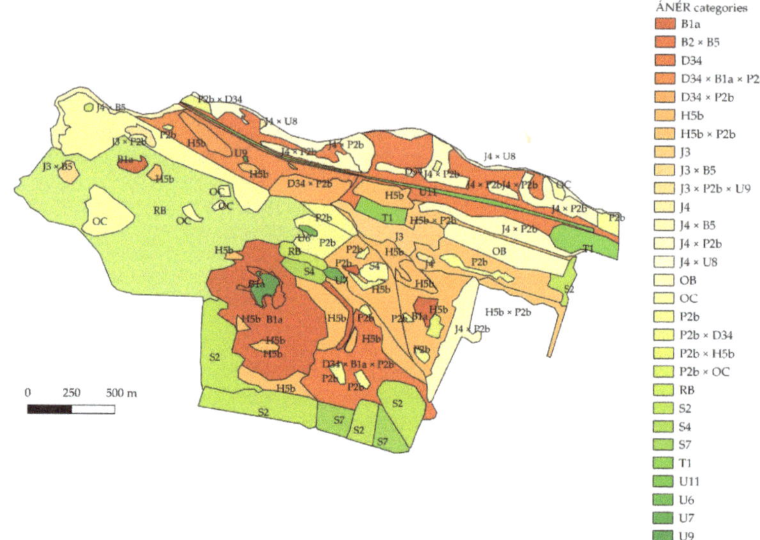

Figure 2. Habitat map of the Dejtár area using ÁNÉR categories.

(A)   **(B)**

Figure 3. Lower seated area of the Ipoly Valley. (**A1**): Gallery forest along the Ipoly; (**A2**): Swamp meadow dominated by *Carex* spp. (**B**): In the belt area in the Ipoly Valley, *Lychnis flos-cuculi* and *Ranunculus* ssp. are in the spring aspect.

On higher elevations, sandy grassland and shrubby, woody patches appear (Figure 4A). The number of patches of sand steppe meadows (H5b) was the highest (16 pieces) and its total area was also large

(63.4 ha). However, considering the associations, they are less continuous than wet patches, because the drier parts of the higher elevations were dominated by *Stipa borysthenica*, which is an element of *Festucion vaginatae* Simon 2000.

Figure 4. (**A**) Sandy grasslands and shrubs appearing in the higher seated parts. (**B**) *Salvia pratensis*, a dominant steppe species. (**C**) *Stipa borysthenica*, a dominant steppe meadow grass species.

In contrast, *Festuca vaginata* W. et K., an important characteristic species of open sandy grasslands is missing and the frequent occurrences of *Salvia pratensis* L. (Lamiaceae) are a sign of the steppe character and causes the vegetation to be akin to *Salvio-nemorosae-Festucetosum rupicolae* Zólyomi ex Soó 1964. From a coenological perspective, it is an interesting situation when a loess-steppe vegetation appears on sandy soil, with *Festuca rupicola* and *Salvia pratensis* as the dominant species (Figure 4B). *Thymus* spp., *Dianthus pontederae* A. Kern. and *Koeleria cristata* (Ledeb.) Schult., which are steppe elements, are also found here. Similarly to the original acidic Pannon sandy grasslands, *Pulsatilla pratensis* ssp. *nigricans* also appears. This vegetation type is very close to *Potentillo arenariae-Festucetum pseudovinae* Soó 1938, 1940, which is a rare association in the eastern part of the Pannonian region, as many of its species appear here.

Furthermore, in the most nutrient-poor parts of the area, diversity of the vegetation of sandy hedges is expanded by *Thymo serpylli-Festucetum pseudovinae* Borhidi 1958 as a new occurrence in the Pannon region. This association was known only from the eastern part of the Pannonian region (Nyírség), and was described in southwestern Hungary (Inner Somogy). Its important characteristic species is *Corynephorus canescens* (L.) P.Beauv. On sandy plains, these habitat patches appear along with pioneer arid and semiarid woody associations (P2b) as a mosaic. This vegetation forms after deforestation or as a consequence of heavy grazing.

3.3. NDVI versus ÁNÉR Categories

Lower vegetation productivity results in lower NDVI value [27]. In the study area (Figure 5), the lowest NDVI value was 0.076 while the highest was 0.83. Urban areas and water bodies have the lowest values, near 0. Still waters (ÁNÉR category: U9) have NDVI values between 0.08 and 0.55. The positive values can be explained by the biological activity in the water but the level of reflection is very low due the chlorophyll-poor areas [30]. Intensively cultivated farmlands (T1) have similar values. The category of roads and railroads (U11) showed low NDVI values similarly to the previous ÁNÉR categories due to the low biological activity; therefore, this category is well separated from the other classes. Swamp meadows (D34) on lower elevations along the river are more unified, while the reefs are distinctly different, which means that the differences originating from the elevation can also be observed in the vegetation of the area. NDVI values are higher where plant activity is higher, or the phenological phase of the plant is in the growing period. More arid sand associations showed low NDVI (0.56–0.632). Most sandy steppe meadows (H5b) along with transitional habitats between grasslands and woody patches had values of 0.632–0.666. Swamp meadow vegetation (D34) is clearly distinguishable from *Saliceto-Populetum* Meijer-Drees 1936 (J4) and dry-semi-dry pioneer shrubs (P2b) complexes along the Ipoly. Woody vegetation did not appear to be very unified, while NDVI categories differed depending on the phenological phase.

Figure 6 shows the different NDVI values of randomly chosen 20 pixels from each ÁNÉR category. Results of the statistical comparison of mean NDVI values among the habitat types are summarized in Table 3. Water bodies (U9) urban areas, such as fallow lands (T1) sand mines (U7), and roads (U11) differed significantly from the majority of other habitats. A somewhat higher but still low NDVI value was found in the sandy steppe meadows (H5b). Woody and grassland vegetation differed from each other. Categories of grassland vegetation, such as sandy steppe meadows (H5b), uncharacteristic arid and semiarid grassland complexes (OC), and uncharacteristic fresh grasslands (OB) showed lower NDVI values. In contrast, woody vegetation patches, such as riverine willow shrubs (J3), planted pinewoods (S4), and Riverine willow-poplar woodlands (J4) significantly differed from most of the non-woody habitats. Dry shrub vegetation with *Crataegus monogyna*, *Prunus spinosa* L. and *Juniperus communis* L. (P2b) showed uniformly medium high values. Wetter and arid grasslands, which are important in terms of grassland farming, differed greatly. Dry grasslands were not uniform because of the species characteristic in them, while in dry grasslands the cover of *Corinephorus canescens* (L.) P.Beauv., *Festuca psedovina* Hack., *Festuca ovina*, *Festuca rupicola*, and *Stipa borysthenica* are larger. These species occur sporadically in the OC category.

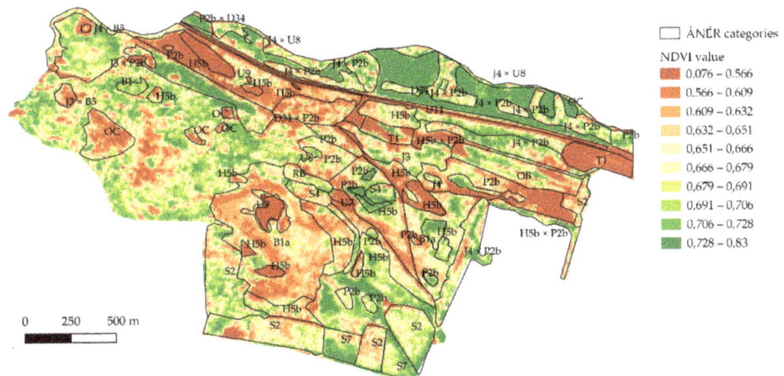

Figure 5. Habitat maps of Dejtár area using ÁNÉR categories and Normalized Vegetation Index (NDVI) data.

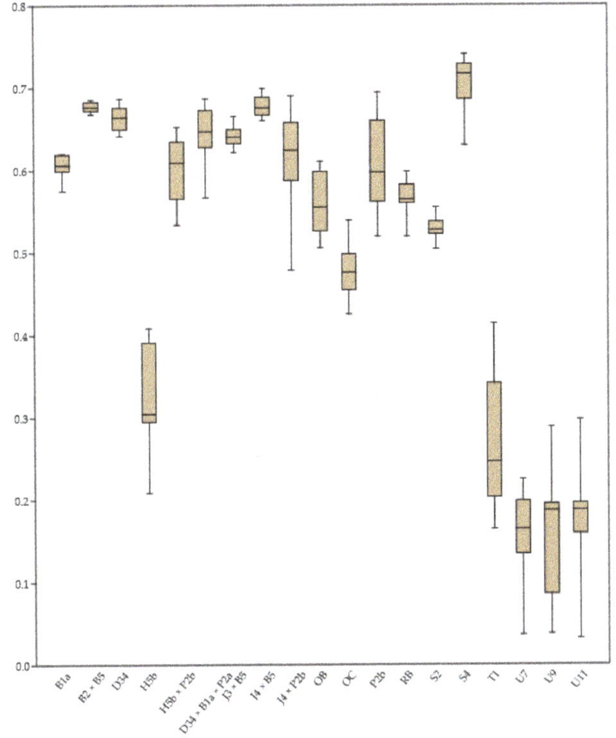

Figure 6. Distribution of ÁNÉR categories (habitat patches) based on NDVI data.

Forests **2021**, 12, 135

Table 3. Statistical comparison of mean NDVI values of the characteristic habitat types in the Ipoly Valley, Hungary. (*: $p < 0.05$, **: $p < 0.01$, ***: $p < 0.001$, ns: not significant).

	B2 × B5	D34	H5b	H5b × P2b	D34 × B1a × P2a	J3 × B5	J4 × B5	J4 × P2b	OB	OC	P2b	RB	S2	S4	T1	U7	U9	U11
B1a	ns	ns	*	ns	ns	ns	ns	ns	ns	ns	ns	ns	ns	*	**	***	***	***
B2 × B5		ns	***	ns	ns	ns	ns	ns	***	***	ns	**	***	ns	***	***	***	***
D34			***	ns	ns	ns	ns	ns	*	***	ns	*	***	ns	***	***	***	***
H5b				*	***	***	***	***	ns	ns	**	ns	ns	***	ns	ns	ns	ns
H5b × P2b					ns	ns	ns	ns	ns	ns	ns	ns	ns	**	**	***	***	***
D34 × B1a × P2a						ns	ns	ns	ns	***	ns	ns	*	ns	***	***	***	***
J3 × B5							ns	ns	ns	**	ns	ns	ns	ns	***	***	***	***
J4 × B5								ns	***	***	ns	**	***	ns	***	***	***	***
J4 × P2b									ns	*	ns	ns	ns	ns	***	***	***	***
OB										ns	ns	ns	ns	***	ns	*	**	*
OC											ns	ns	ns	***	ns	ns	ns	ns
P2b												ns	ns	ns	***	***	***	***
RB													ns	***	ns	**	**	**
S2														***	ns	ns	ns	ns
S4															***	***	***	***
T1																ns	ns	ns
U7																	ns	ns
U9																		ns

4. Discussion

In the present vegetation study, we also discovered two vegetation types: the *Potentillo arenariae-Festucetum pseudovinae* and the *Thymo serpylli-Festucetum pseudovinae*. The two vegetation types were discovered in the Pannonian region as a new occurrence, which may have appeared as a result of environmental conditions [8,31]. The sand hills protruding from the river floodplain offered dry and nutrient-poor habitats where sand vegetation appeared and was able to form.

Vegetation edges were studied primarily in the central Great Plain of the Carpathian Basin [12,14]. Similar transitional conditions under similar environmental, water, and soil conditions may develop at the northern occurrence limits of these typical Pannonian vegetation types. In addition, two vegetation types were discovered in the Pannonian region as a new occurrence, which may have appeared as a result of territorial utilization, as in the eastern part of the basin (Nyírség) [31]. During the examination of the sandy area, two new occurrences of acerbic sand were found in the Pannonian region, which were only known from the western (Inner Plain), southwestern (Inner Somogy), and eastern (Nyírség) areas of the Pannonian region [36,37].

Among the dominant species such as *Corinophorus canescens* (L.) P. B., *Jasione montana* L., *Veronica* ssp.; however, the dominant species *Festuca vaginata* did not occur. The dominant taxon in the more closed stands was *Festuca pseudovina*, which also indicates the degradation of the vegetation [37–39]. The occurrence of *Festuca rupicola* suggests cooler environmental conditions and climatic effects [40].

Complex patches appear in the deeper areas, mostly characteristic of individual Hungarian habitats [22, 41,42], and wet, swampy, marshy, or water-bound vegetation patches appear in the depressions on the site, which [43–45] may generally be typical but the position of the ground water table is especially important in the central area of the Carpathian Basin, which leads to the appearance of diverse and species rich vegetation [15,16,45]. Spots from the ÁNÉR-based habitat mapping showed agreement with the remote sensing data, which were also used as controls. Typical sandy grassland species such as *Stipa borysthenica* also occur at the heights of the northern sand steppes. In addition, isolated parts within each patch can be detected, which provide additional information about land use, which is also of practical importance for grazing, where land use is important in its maintenance, as well as in the mosaic-like and diverse vegetation.

Data from the Sentinel-2A satellite offer an opportunity for mapping natural habitats [46–48]. Based on our observations, it can be seen that the individual vegetation patches can be distinguished well based on the ÁNÉR categories, thus the field mapping is facilitated by the data of satellite images in hard-to-reach areas, as others noticed during their work [27]. When constructing an association-accurate map, the individual vegetation types are not clearly separated if there is no need for a habitat map of such accuracy and a simpler category system can be used to solve this problem, which [27] also received attention during the study. In contrast, the regularities are clearly outlined in the studied habitats and the agricultural area shows a homogeneous picture since after the harvest the open farmland (Category T1 in Figure 5) shows a lower NDVI value [49,50]. One can see that the wet and dry spots are markedly different in between the connected areas. The sandy grasslands are different associations on the bedrock, giving a mosaic-like picture (Category H5b in Figure 5) [46,51]. Isolated areas within each patch can also be identified as an area used by cattle as resting areas to provide additional information on land use that is also of practical importance for grazing. In order to preserve the original vegetation of the area [52], we need to protect the endangered and protected plants and also the vegetation types from river regulation, ploughing, and deforestation of the areas.

5. Conclusions

Answering question (i), the northern boundary of the Pannonian steppe-forest-steppe vegetation type is found in the study area, which is the valley of the Ipoly River. The vegetation type appears in the drier, acerbic sandy areas along the rivers, as well as new occurrence patches, and a special coenosystemic vegetation mixture is formed. (ii) In the study area, the nearby sand ridges provided an environmental background to learn about the boundaries of the Pannonian sand steppe and forest-steppe. The dominant species of the steppe patches was *Festuca rupicola*, which plays a similar coenosystemic role in the formation of the vegetation to *Festuca vaginata* in the central parts of the Carpathian Basin. Based on the data, the appearance of *Festuca vaginata* can also be expected due to global climatic changes [53–55]. (iii) Based on the Sentinel-2A data, we have seen that urban areas are separated from natural habitats but the individual vegetation types are not clearly separated.

Author Contributions: Conceptualization, I.J., F.S., and K.P.; Data curation, E.S.-F. and N.P.; Formal analysis, Z.Z.; Methodology, I.J., D.F., and P.C.; Software, D.S.; Supervision, K.P.; Writing—original draft, I.J.; Writing—review & editing, G.P. and K.P. All authors have read and agreed to the published version of the manuscript.

Funding: The work was funded by OTKA K-125423 and NKFIH-1159-6/2019.

Institutional Review Board Statement: Not applicable.

Informed Consent Statement: Not applicable.

Data Availability Statement: The data presented in this study are available on request from the corresponding author.

Acknowledgments: Our studies received financial support from the Higher Education Institutional Excellence Program (NKFIH-1159-6/2019) awarded by the Ministry for Innovation and Technology and by OTKA K-125423.

Conflicts of Interest: The authors declare no conflict of interest.

References

1. Török, P.; Penksza, K.; Tóth, E.; Kelemen, A.; Sonkoly, J.; Tóthmérész, B. Vegetation type and grazing intensity jointly shape grasing on grassland biodiversity. *Ecol. Evol.* **2018**, *8*, 10326–10335. [CrossRef]
2. Wesche, K.; Ambarlı, D.; Kamp, J.; Török, P.; Treiber, J.; Dengler, J. The Palaearctic steppe biome: A new synthesis. *Biodivers. Conserv.* **2016**, *25*, 2197–2231. [CrossRef]
3. Zólyomi, B. *Budapest és Környékének Természetes Növénytakarója*; Pécsi, M., Ed.; Budapest Természeti képe; Akadémiai Kiadó: Budapest, Hungary, 1958.
4. Szujkóné-Lacza, J.; Kováts, D. *The Flora of the Kiskunság National Park*; Magyar Természettudományi Múzeum: Budapest, Hungary, 1993; p. 469.
5. Kárpáti, I.; Kárpáti, V. The aspects of the calciphilous turf (Festucetum vaginale danubiale) in the environs of Vácrátót in 1952. *Acta Bot. Hung.* **1954**, *1*, 129–157.
6. Fekete, G.; Tuba, Z.; Melkó, E. Background processes at the population level during succession in grasslands on sand. *Vegetatio* **1988**, *77*, 33–41. [CrossRef]
7. Varga, Z. Die Waldteppen des pannonischen Raumes aus biogeographischer Sicht. *Düsseldorfer Geobotanisches Kolloquium* **1989**, *6*, 35–50.
8. Molnár, Z.; Biró, M.; Bartha, S.; Fekete, G. Past trends, present state and future prospects of Hungarian forest-steppes. In *Ecological Problems and Livelihoods in a Changing World*; Werger, M.J.A., van Staalduinen, M.A., Eds.; Springer: Dordrecht, The Netherlands; Heidelberg, Germany; New York, NY, USA; London, UK, 2012; pp. 209–252.
9. Varga, Z.; Borhidi, A.; Fekete, G.; Debreczy, Z.; Bartha, D.; Bölöni, J.; Molnár, A.; Kun, A.; Molnár, Z.; Lendvai, G.; et al. *Alföldi Erdőssztyeppmaradványok Magyarországon/Forest Steppe Remains in the Great Plain of Hungary*; WWF-MTA ÖBKI: Budapest-Vácrátót, Hungary, 2000.
10. Rédei, T.; Csecserits, A.; Barabás, S.; Lhotsky, B.; Botta-Dukát, Z. Homoki erdőssztyeppmozaikok kiterjedésének és változatosságának hatása a fajgazdagságra. The effect of size mand habitat diversity of sand forest steppe mosaics on plant species richness. *Természetvédelem és Kutatás a Turjánvidék Északi Részén. Rosalia* **2018**, *10*, 131–144.
11. Magyari, E.K.; Chapman, J.C.; Passmore, D.G.; Allen, J.R.M.; Huntley, J.P.; Huntley, B. Holocene persistence of wooded steppe in the Great Hungarian Plain. *J. Biogeogr.* **2010**, *37*, 915–935. [CrossRef]
12. Biró, M.; Czúcz, B.; Horváth, F.; Révész, A.; Csatári, B.; Molnár, Z. Drivers of grassland loss in Hungary during the post-socialist transformation (1987–1999). *Landsc. Ecol.* **2013**, *28*, 789–803. [CrossRef]
13. Biró, M.; Horváth, F.; Révész, A.; Molnár, Z.; Vajda, Z. Száraz homoki élőhelyek és átalakulásuk a Duna–Tisza közén a 18. századtól napjainkig. In *Természetvédelem és Kutatás a Duna–Tisza közi Homokhátságon. Rosalia 6. Duna–Ipoly Nemzeti Park Igazgatóság*; Verő, G., Ed.; Duna Ipoly Nemzeti Park Igazgatóság: Budapest, Hungary, 2011; pp. 383–421.
14. Tölgyesi, C.; Körmöczi, L. Structural changes of a Pannonian grassland plant community in relation to the decrease of water availability. *Acta Bot. Hung.* **2012**, *54*, 413–431. [CrossRef]
15. Tölgyesi, C.; Bátori, Z.; Erdős, L. Using statistical tests on relative ecological indicator values to compare vegetation units–Different approaches and weighting methods. *Ecol. Indic.* **2014**, *36*, 441–446. [CrossRef]
16. Erdős, L.; Tölgyesi, C.; Horzse, M.; Tolnay, D.; Hurton, Á.; Schulcz, N.; Körmöczi, L.; Lengyel, A.; Bátori, Z. Habitat complexity of the Pannonian forest-steppe zone and its nature conservation implications. *Ecol. Complex.* **2014**, *17*, 107–118. [CrossRef]
17. Erdős, L.; Kröel-Dulay, G.; Bátori, Z.; Kovács, B.; Németh, C.; Kiss, P.J.; Tölgyesi, C. Habitat heterogeneity as a key to high conservation value in forest-grassland mosaics. *Biol. Conserv.* **2018**, *226*, 72–80. [CrossRef]

18. Erdős, L.; Török, P.; Szitár, K.; Bátori, Z.; Tölgyesi, C.; Kiss, P.J.; Bede-Fazekas, Á.; Kröel-Dulay, G. Beyond the forest-grassland dichotomy: The gradient-like organization of habitats in forest-steppes. *Front. Plant Sci.* **2020**, *11*, 236. [CrossRef]

19. Tölgyesi, C.; Valkó, O.; Deák, B.; Kelemen, A.; Bragina, T.M.; Gallé, R.; Erdős, L.; Bátori, Z. Tree–herb co-existence and community assembly in natural forest-steppe transitions. *Plant Ecol. Divers.* **2018**, *11*, 465–477. [CrossRef]

20. Fodor, I.; Gálosi-Kovács, B. A Kárpát-medence határokon átnyúló természeti értékei. In *A környezet és a Határok Kutatója. Tiszteletkötet Nagy Imre 65. Születésnapja Alkalmából, Regionális Tudományi Társaság*; Gál, Z., Ricz, A., Eds.; pp. 39–51, ISBN 978-86-86929-07-5. Available online: http://hdl.handle.net/11155/2088 (accessed on 19 December 2019).

21. Brow, A.G.; Lespez, L.; ASear, D.; Macaire, J.; Houben, P.; Klimek, K.; Brazier, R.E.; Van Oost, K.; Pears, B. Natural vs. anthropogenic streams in Europe: History, ecology and implications for restoration, river-rewilding and riverine ecosystem services. *Earth-Sci. Rev.* **2018**, *180*, 185–205. [CrossRef]

22. Penksza, K.; Nagy, A.; Laborczi, A.; Pintér, B.; Házi, J. Wet habitats along River Ipoly (Hungary) in 2000 (extremely dry) and 2010 (extremely wet). *J. Maps* **2012**, *8*, 157–164. [CrossRef]

23. Mjazovszky, Á.; Csontos, P.; Tamás, J. A patakkísérő növényzet vizsgálata négy hazai táj viszonylatában. *Botanikai Közlemények* **2007**, *94*, 45–55.

24. Verrasztó, Z. Környezeti monitoring vizsgálatok az Ipoly vízgyűjtőjén (célkitűzések és általános tájékoztatás). *Tájökológiai Lapok* **2010**, *8*, 535–561.

25. Járdi, I.; Pápay, G.; Fekete, G.; S.-Falusi, E. Marhalegelők vegetációjának vizsgálata az Ipoly-völgy homoki gyepeiben. *Gyepgazdálkodási Közlemények* **2017**, *15*, 9–21.

26. Sentinel-2A Website. Available online: https://scihub.copernicus.eu/ (accessed on 23 June 2015).

27. Burai, P.; Lenart, C.; Valkó, O.; Bekő, L.; Szabó, Z.; Deák, B. Fátlan vegetációtípusok azonosítása légi hiperspektrális távérzékelési módszerrel. *Tájökológiai Lapok* **2016**, *14*, 1–12.

28. Marosi, S.; Somogyi, S. *Magyarország Kistájainak Katasztere I-II*; MTA Földrajztudományi Kutató Intézet: Budapest, Hungary, 1990; 1023p.

29. Fekete, G.; Molnár, Z.; Horváth, F. (Eds.) *A Magyarországi Élőhelyek Leírása, Határozója és a Nemzeti Élőhely-osztályozási Rendszer. (Description of and Guide to Hungarian Habitats)*; Magyar Természettudományi Múzeum: Budapest, Hungary, 1997.

30. Bölöni, J.; Molnár, Z.; Horváth, F.; Illyés, E. Naturalness-based habitat quality of the hungarian (semi-) natural habitats. *Acta Bot. Hung.* **2008**, *50*, 149–159. [CrossRef]

31. Bölöni, J.; Molnár, Z.; Kun, A. (Eds.) *Magyarország Élőhelyei. A Hazai Vegetációtípusok Leírása és Határozója*; ÁNÉR 2011; MTA ÖBKI: Vácrátót, Hungary, 2011; p. 441.

32. Borhidi, A.; Kevey, B.; Lendvai, G.; Seregélyes, T. *Plant Communities of Hungary*; Akadémiai Kiadó: Budapest, Hungary, 2012.

33. Király, G. (Ed.) *Új Magyar Füvészkönyv. Magyarország Hajtásos Növényei. Határozókulcsok [New Hungarian Herbal. The Vascular Plants of Hungary. Identification Key]*; ANP Igazgatóság: Jósvafő, Hungary, 2009.

34. Didan, D. *MOD13Q1 MODIS/Terra Vegetation Indices 16-Day L3 Global 250m SIN Grid V006*; NASA EOSDIS Land Processes DAAC; NASA: College Park, MD, USA, 2015.

35. Hammer, Ø.; Harper, D.A.T.; Paul, D.R. Past: Paleontological Statistics Software Package for Education and Data Analysis. *Palaeontol. Electr.* **2001**, *4*, 9.

36. Borhidi, A. *Magyarország Növénytárulásai*; Akadémiai Kiadó: Budapest, Hungary, 2003; p. 610.

37. Borhidi, A. Social behaviour types, the naturalness and relative ecological indicator values of the higher plants in the Hungarian flora. *Acta Bot. Acad. Sci. Hung.* **1995**, *39*, 97–181.

38. Simon, T. *A Magyarországi Edényes Flóra Határozója*; Tankönyvkiadó: Budapest, Hungary, 2000.

39. Soó, R. *A Magyar Flóra és Vegetáció Rendszertani-Növényföldrajzi Kézikönyve I. (Synopsis Systematico-Geobotanica Florae Vegetationisque Hungariae I)*; Akadémia Kiadó: Budapest, Hungary, 1964.

40. Pongrácz, R.; Bartholy, J.; Szabó, P.; Gelybó, G. A comparison of observed trends and simulated changes in extreme climate indices in the Carpathian Basin by the end of this century. *Int. J. Glob. Warm.* **2009**, *1*, 336–355. [CrossRef]

41. Malatinszky, Á.; Ádám, S.; Falusi, E.; Saláta, D.; Penksza, K. Climate change related land user problems in protected wetlands: A study in a seriously affected Hungarian area. *Clim. Chang.* **2013**, *118*, 671–683. [CrossRef]
42. Uj, B.; Nagy, A.; Saláta, D.; Laborczi, A.; Malatinszky, A.; Bakó, G.; Danyik, T.; Tóth, A.; Falusi, E.S.; Gyuricza, C.; et al. Habitat map of Kis-Sárrét (Körös-Maros National Park, Hungary) with special regard to the changes of wetlands. *J. Maps* **2015**, *10*, 211–221. [CrossRef]
43. Körner, C. A reassessment of high-elevation treeline positions and their explanation. *Oecologia* **1998**, *115*, 445–459.
44. Courtwright, J.; Findlay, S.E.G. Effects of microtopography on hydrology, physicochemistry, and vegetation in a tidal swamp of the Hudson River. *Wetlands* **2011**, *31*, 239–249. [CrossRef]
45. Bátori, Z.; Farkas, T.; Erdős, L.; Tölgyesi, C.; Körmöczi, L.; Vojtkó, A. A comparison of the vegetation of forested and non-forested solution dolines in Hungary: A preliminary study. *Biologia* **2014**, *69*, 1339–1348. [CrossRef]
46. Bekkema, M.; Eleveld, M. Mapping Grassland Management Intensity Using Sentinel-2 Satellite Data. *GI_Forum* **2018**, *1*, 194–213. [CrossRef]
47. Kaplan, G. Mapping and Monitoring Wetlands Using Sentinel 2 Satellite Imagery. Available online: https://pdfs.semanticscholar.org/a101/515a9d639c896364cec0b589172af3649717.pdf (accessed on 15 October 2017).
48. Majasalmi, T.; Rautiainen, M. The potential of Sentinel-2 data for estimating biophysical variables in a boreal forest: A simulation study. *Remote Sens. Lett.* **2016**, *7*, 427–436. [CrossRef]
49. Veloso, A.; Stéphane, M.; Bouvet, A.; Le Toan, T.; Planells, M.; Dejoux, J.; Ceschia, E. Understanding the temporal behavior of crops using Sentinel-1 and Sentinel-2-like data for agricultural applications. *Remote Sens. Environ.* **2017**, *199*, 415–426. [CrossRef]
50. Beck, P.S.A.; Jönsson, P.; Høgda, K.-A.; Karlsen, S.R.; Eklundh, L.; Skidmore, A.K. A ground-validated NDVI dataset for monitoring vegetation dynamics and mapping phenology in Fennoscandia and the Kola peninsula. *Int. J. Remote Sens.* **2007**, *28*, 4311–4330. [CrossRef]
51. Bekkema, M.E. The Potential of Sentinel-2 Data for Detecting Grassland Management Intensity to Support Monitoring of Meadow Bird Populations. Master's Thesis, UNIGIS, Amsterdam, The Netherlands, 2017.
52. Maglocký, Š.; Feráková, V. Red list of ferns and flowering plants (*Pteridophyta* and *Spermathophyta*) of the flora of Slovakia. *Biologia* **1993**, *48*, 361–385.
53. Zsilinszki, A.; Dezső, Z.; Bartholy, J.; Pongrácz, R. Synoptic-climatological analysis of high level air flow over the Carpathian Basin. *Időjárás/Q. J. Hung. Meteorol. Serv.* **2019**, *123*, 19–38. [CrossRef]
54. Bartholy, J.; Pongrácz, R. Tendencies of extreme climate indicates based on daily precipitation in the Carpathian Basin for the 20th century. *Időjárás* **2005**, *109*, 1–20.
55. Bartholy, J.; Pongrácz, R. Comparing tendencies of some temperature and precipitation on global and regional scales. *Időjárás* **2006**, *110*, 35–48.

Publisher's Note: MDPI stays neutral with regard to jurisdictional claims in published maps and institutional affiliations.

Article

Does Shrub Encroachment Indicate Ecosystem Degradation? A Perspective Based on the Spatial Patterns of Woody Plants in a Temperate Savanna-Like Ecosystem of Inner Mongolia, China

Xiao Wang [1], Lina Jiang [2], Xiaohui Yang [3], Zhongjie Shi [3,*] and Pengtao Yu [1]

[1] Key Laboratory of Forest Ecology and Environment of the State Forestry Administration of China, Research Institute of Forest Ecology, Environment and Protection, Chinese Academy of Forestry, Beijing 100091, China; wangxiao@caf.ac.cn (X.W.); yupt@caf.ac.cn (P.Y.)

[2] Research Institute of Forestry New Technology, Chinese Academy of Forestry, Beijing 100091, China; jianglina@caf.ac.cn

[3] Institute of Desertification Studies, Chinese Academy of Forestry, Beijing 100091, China; yangxh@caf.ac.cn

* Correspondence: shizj@caf.ac.cn; Tel.: +86-10-6282-4106

Received: 7 October 2020; Accepted: 23 November 2020; Published: 25 November 2020

Abstract: Shrub encroachment, i.e., shrub emergence or an increase in woody plant cover, has been widely observed in arid and semiarid grasslands and savannas worldwide since the 2000s. However, until now, there has been a clear division of opinion regarding its ecological implications. One view is that shrub encroachment is an indicator of ecological degradation, and the other is that shrub encroachment is a sign of the restoration of degraded ecosystems. This division leads to completely different judgments about the states and transition phases of shrub-encroached ecosystems, which further affects decisions about their conservation and management. To determine whether ecosystems experiencing shrub encroachment are degrading or are in a postdegraded restoration stage, the spatial distributions and interactions of woody plants after shrub encroachment were investigated in this study. An *Ulmus pumila*-dominated temperate savanna-like ecosystem with significant shrub encroachment in the Otindag Sandy Land, Inner Mongolia, China, was selected as the research area, and woody plants were surveyed within a 25-hectare (500 × 500 m) plot. Spatial point pattern analysis was employed to analyze the distribution patterns of the woody plants. The results indicated different patterns for *U. pumila* trees, i.e., a random distribution pattern for old trees (with a diameter at breast height (DBH) of more than 20 cm) and aggregated distribution patterns for medium (5 cm ≤ DBH < 20 cm) and juvenile trees (DBH < 5 cm) at scales of 0–9 and 0–12 m, respectively. For most shrubs, there was significant aggregation at a scale of 0–6 m. However, there were significant negative relationships between old *U. pumila* trees (DBH ≥ 20 cm) and most shrub species, such as *Caragana microphylla* and *Spiraea aquilegifolia*. In contrast, there were positive relationships between juvenile trees (DBH < 5 cm) and most shrub species. These results suggest that, to some extent, shrub encroachment may have disrupted the normal succession pattern in the *U. pumila* community in this area, and without conservation, the original tree-dominated temperate savanna-like ecosystem may continue to deteriorate and eventually become a shrub-dominated temperate savanna-like ecosystem.

Keywords: shrub encroachment; spatial pattern; temperate savanna; ecosystem degradation

---

1. Introduction

Shrub encroachment in grasslands and the densification of woody plant cover in savannas have been widely documented across many arid and semiarid areas of the world [1–4], including in

South America, Australia, and the warm deserts of the Southwestern United States. However, until now, there has been a clear division of opinion about its ecological indication. One view is that shrub encroachment is an indicator of land degradation that is often associated with ecosystem degradation [5], such as declines in forage productivity, biodiversity, and socioeconomic potential, as well as increased erosion [6]. However, an alternative viewpoint proposing that shrub emergence is a sign of the restoration of degraded ecosystems has emerged recently [7,8], considered to support the biodiversity and a variety of ecosystem services [9]. In addition, shrub encroachment has been described as an alternative stable state occurring several times during the last two millennia in African savannas [10]. Some studies also found that the role of shrub species was important in the rehabilitation of degraded sandy land ecosystems [11]. Furthermore, different opinions exist as to the causes of shrub encroachment in different types of ecosystems [12]: some studies found that underuse leads to shrub and subsequent tree encroachment and, finally, conversion to forest [12]; however, overgrazing usually causes shrub encroachment in Africa and the Mongolia Plateau [10]. This difference in the perception of shrub encroachment could lead to completely different judgments of the states and transitions of shrub-encroached ecosystems, which would further affect decisions about their conservation and management [13]. An innovative study [14], combined with the landscape history and other environmental factors, quantified the process and indicative significance of shrub encroachment and provided us with new ideas. Therefore, more attention needs to be paid to ecosystem changes after shrub encroachment to clarify their ecological significance, especially the significance of the degree of shrub encroachment and changes in distribution patterns and structure. This information could then be used to quantify the ecological indications of shrub encroachment.

As a common tree species distributed widely throughout the forest-steppe ecotone on the Mongolian Plateau [15,16], *Ulmus pumila* forms a stable savanna-like woody-herbaceous complex ecosystem in association with grass in the Horqin Sandy Land, the Otindag Sandy Land, and the Hulunbuir Sandy Land in Northern China [17,18]. The sparse *U. pumila* trees on the temperate savanna-like ecosystem have ecological significance because they provide sand stabilization and also resting places for livestock [19]. In recent years, due to various changing conditions, such as climate change and land use change in the course of economic development [20–22], there has been severe damage to the *U. pumila*-dominated temperate savanna-like ecosystem. This damage has led to an increase in the number of shrubs and the densification of woody plant cover as well as outcomes such as decreasing temperate savanna-like ecosystem area, the loss of structural integrity, poor population regeneration, and changing spatial patterns [23,24]. Previous studies have focused mainly on the vegetation distribution, possible encroachment mechanisms, vegetation characteristics, species composition, and carbon budget in the *U. pumila*-dominated temperate savanna-like ecosystem after shrub encroachment [25,26]. It was found that shrub coverage in the grasslands of Inner Mongolia reached 12.8% and that the emergence of *Caragana microphylla* was an indicator of grassland degradation in Inner Mongolia [27]. However, the spatial patterns of woody plants and their interrelationships associated with shrub encroachment are often overlooked despite having important effects on ecosystem functions [28].

Spatial pattern analysis is an important method for studying the distribution and relationships of different plants [29,30]. The spatial patterns of species and the spatial relationships among species significantly impact growth, reproduction, death, resource utilization, and gap formation among species [31,32]. The analysis of a species' spatial pattern helps us to understand both the ecological processes that form the pattern (such as intra- and interspecific competition, interference, and environmental heterogeneity) and the ecophysiological traits of the plant species, including the relationships between these plants and their environment [33,34]. Recently, a spatial pattern analysis method has been used to clarify the vegetation degradation processes underlying the patterns of individual species in semiarid and arid areas [35,36].

In this study, our objective is to explore how shrub encroachment affects the spatial distribution and interaction of woody plants in a temperate savanna-like ecosystem. We hypothesize that there are differences in the relationships between *U. pumila* trees of different DBHs and shrubs. The following questions will be addressed: (1) What are the distribution patterns of *U. pumila* trees and shrubs in a temperate savanna-like ecosystem? (2) What are the interspecific interactions between *U. pumila* trees and shrubs? and (3) Do tree–shrub spatial associations change as a function of tree size?

2. Materials and Methods

2.1. Study Area

U. pumila-dominated temperate savanna-like ecosystem, which is distributed widely throughout the forest-steppe ecotone on the Mongolian Plateau, China, is quite different from North American prairies and African savannas in terms of climate, soils, and dominant plant functional types; however, there are some taxonomical similarities among these ecosystems at the genus and family levels [37,38]. The study area experiences a typical temperate continental semiarid climate, with warm summers and cold winters. The mean annual precipitation is approximately 350 mm, 70% of which occurs from June to August. The mean annual pan evaporation is 1900 mm. The mean annual air temperature is 1.6 °C, with a maximum monthly temperature of 17.8 °C (July) and a minimum monthly temperature of −17.6 °C (January) [37,38]. The main soil types are aeolian sandy soils with a mean thickness of 200 cm and a calcic horizon occurring at 30–100 cm depth. In this forest-steppe ecotone, *U. pumila* is dominant; the other species that occur include other woody species, including *C. microphylla*, *Spiraea aquilegifolia*, *Ribes diacanthum* and *Salix linearistipularis*, and herbaceous plants, including *Leymus chinense* and *Cleistogenes squarrosa* [39]. The main land use in the area is grazing, and the livestock are primarily cattle. The local herders have implemented alternating seasonal grazing (winter and spring) and mowing (autumn) regimes since 2000 [40]; before that time, annual grazing was the local practice. Shrub encroachment, such as the spread of *C. microphylla*, has been clearly observed in grasslands and savannas in this area, resulting in a landscape characterized by a mosaic of shrub and grass patches.

2.2. Research Site and Data Collection

Most work about the spatial distribution and interaction of woody plants had been done in single hectare or smaller plots, but the relative rarity of many species in forests necessitated large-scale census plots. Thus, plots usually with more than 5 hectares, named large plots, are considered as representative of local vegetation, which could cover the local typical vegetation and topography in a region. By a large plot, the interference of scale and environmental heterogeneity could be avoided for the study on vegetation composition, pattern, and biodiversity [41]. This method has been widely used in global forest dynamic monitoring.

In this study, this large plot method was applied and one large plot with an area of 25 ha (500 m × 500 m) was established following the plot standards of the Center for Tropical Forest Science (CTFS) network [41]. It was located in the Otindag Sandy Land in Inner Mongolia, China (115°16′ E, 42°50′ N), in a typical area of *Ulmus pumila*-dominated temperate savanna-like ecosystem (Figure 1).

Then, this large plot was further divided into 625 subplots (20 m × 20 m). All free-standing trees with stem diameters at breast height (DBHs) of more than one centimeter and all shrubs were tagged, mapped, and identified to species during the summer of 2013–2014. The coordinates of woody plants in each subplot were recorded using an Electronic Total Station, with the southwestern corner of the subplot as the origin. Additionally, in the center of each subplot, a small plot with the size of 1 × 1 m was setup. All herbaceous species were identified and their cover ratios and numbers were recorded.

All *U. pumila* trees were classified into three categories according to their DBH [42], namely, old trees (DBH ≥ 20 cm), medium trees (5 cm ≤ DBH < 20 cm), and juvenile trees (DBH < 5 cm) (Table 1).

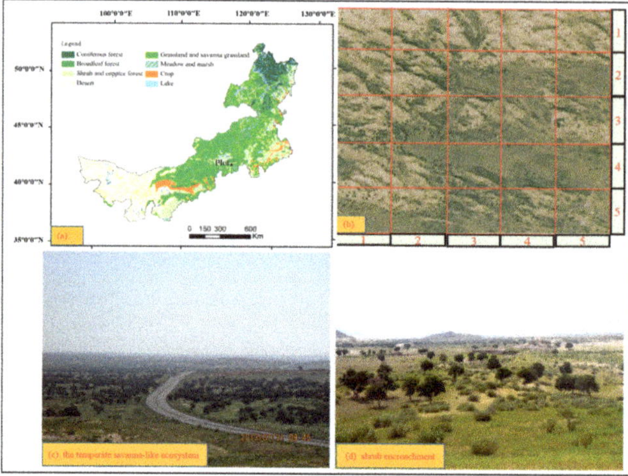

Figure 1. Map of the Otindag Sandy Land and the location of the study region (**a**) location; (**b**) sample map; (**c,d**) pictures of the landscape.

Table 1. Basic parameters of *U. pumila* trees and shrubs.

Species	Category	Number	Height (m)	Maximum Crown Radius (m)	Minimum Crown Radius (m)
U. pumila trees	Old	750	8.4 ± 0.94	6.2 ± 0.83	5.0 ± 1.53
	Medium	250	5.3 ± 0.93	4.1 ± 0.25	3.2 ± 1.27
	Juvenile	400	1.7 ± 0.30	1.5 ± 0.30	0.4 ± 0.22
Shrubs	*Salix linearistipularis*	292	1.0 ± 0.13	1.1 ± 0.28	0.4 ± 0.10
	S. microstachya var. *bordensis*	547	0.50 ± 0.17	0.4 ± 0.15	0.20 ± 0.35
	S. gordejevii	238	1.30 ± 0.22	0.6 ± 0.11	0.40 ± 0.27
	S. myrtilloides	837	1.20 ± 0.31	0.7 ± 0.13	0.4 ± 0.34
	Betula fruticosa	2417	1.0 ± 0.14	0.5 ± 0.18	0.3 ± 0.10
	Ribes diacanthum	168	0.8 ± 0.16	0.6 ± 0.23	0.1 ± 0.11
	Caragana microphylla	414	0.3 ± 0.12	0.4 ± 0.23	0.35 ± 0.14
	Spiraea aquilegifolia	1038	0.6 ± 0.15	0.8 ± 0.14	0.6 ± 0.12

2.3. Data Analysis

The pair-correlation function g (r) is a statistical method used to estimate the number of points within concentric rings at a distance r rather than within a certain radius and is especially sensitive to small-scale effects [43,44]. There are two g (r) functions, i.e., univariate and bivariate g (r). In this study, the univariate g (r) function was used to analyze the spatial distribution patterns within woody plants, and the bivariate g (r) function to quantify the both intra- and interspecific spatial association among tree and shrub plants [43,44].

Firstly, the univariate g (r) function was used to analyze three tree categories (old trees, medium trees, and juvenile trees) and all shrub species. The null model of complete spatial randomness (CSR) as a null hypothesis was used for all the univariate analyses. Secondly, for the bivariate analyses, two cases were considered. One case was that the relationship between small and large trees was considered. Since large trees may impact the distribution pattern of small trees within their area of influence (competition), a bivariate g function analysis was conducted for these two size classes using both the toroidal shift and the antecedent condition null model options [43]. This tests whether the patterns of distribution of small and large trees were generated by independent processes. The antecedent condition model tests whether one pattern (small trees) is influenced by a second pattern (large trees), assessing whether there are more (or fewer) small trees in the neighborhood of large trees than expected under a random distribution of small trees [43]. The second case concerns the interaction between trees and shrubs. Because the spatial distributions of plants in plots seem to be affected significantly by drought stress and habitat heterogeneity (e.g., soil patch and microtopography), we examined the spatial association between the two species with the independent null model [44].

All analyses were conducted with Programita software (2016). To assess the significance of the data by comparing them with the null model, 95% confidence intervals were obtained by running 199 Monte Carlo simulations [45].

3. Results

3.1. Characteristics of the Temperate Savanna Ecosystem

There were nine woody plant species, *S. linearistipularis*, *S. microstachya var. bordensis*, *S. gordejevii*, *S. myrtilloides*, *B. fruticosa*, *C. microphylla*, *Spiraea aquilegifolia*, *R. diacanthum*, and *U. pumila*, which were identified in the plot. The trees in the plot mainly grew as individual trees rather than in groups (Table 1).

U. pumila was the dominant tree species, with a coverage of 10–15% and a density of 56 trees/ha. The densities of old trees (DBH ≥ 20 cm), medium trees (5 cm ≤ DBH < 20 cm), and juvenile trees (DBH < 5 cm) were 30, 10, and 16 trees/ha, respectively. The tree heights of these three categories varied, i.e., old trees, medium trees, and juvenile trees had average heights of 8.4 ± 0.94, 5.3 ± 0.93, and 1.7 ± 0.30 m, respectively. Although the coverage of the shrub layer was only 8–15%, eight shrub species were observed, and their density varied from 6.7 to 96.7 plants/ha; the shrubs were shorter than 0.3–1.3 m on average. These shrubs normally grew in clusters distributed in the canopy gaps among *U. pumila* trees.

The herbs were also unevenly distributed and were rarely found growing under the old trees. However, the coverage of the herb layer, 30–60%, was greater than that of the tree and shrub layers. The herbs were also very diverse, with a total of 195 species from 40 families identified in our research plot, representing approximately 30% of the total number of higher plants observed in the Otindag Sandy Land [18,39]. The herbs mainly belonged to the *Compositae* (38 species), *Gramineae* (30 species), and *Leguminosae* (14 species) families.

3.2. Spatial Distribution Patterns of U. pumila Trees—A Random Distribution of Old Trees and an Aggregated Distribution of Juvenile Trees

As shown in Figure 2, *U. pumila* trees showed significant aggregation at the 0–7 m scale. However, their distribution patterns varied with tree size. The old *U. pumila* trees were randomly distributed at all scales (0–50 m); in contrast, the medium *U. pumila* trees (5 cm ≤ DBH < 20 cm) exhibited significant aggregation at the 0–5 m scale, and juvenile trees (DBH < 5 cm) were significantly aggregated at the 0–9 m scale. However, all three tree categories were randomly distributed at other scales in the 25-ha permanent plot.

As shown in Figure 3, there were intraspecific relationships between the different DBH classes of *U. pumila* trees. There was a significant negative relationship between old trees and medium trees at the 2–10 m scale, but no obvious relationships between old trees and juvenile trees. On the other hand, there was a significantly positive relationship between medium trees and juvenile trees at the 2–12 m scale, but no significant relationship at any other scale.

3.3. Aggregated Distribution of Shrubs at Specific Scales

As shown in Figure 4, most shrub species exhibited spatial aggregation at small scales. For example, *S. aquilegifolia*, *S. microstachya* var. *bordensis*, *S. gordejevii*, and *C. microphylla* were significantly aggregated at scales of 0–6, 1–7, 0–2, and 0–6 m, respectively. However, they exhibited random distributions at other scales. *S. myrtilloides* and *B. fruticosa* exhibited similar distribution patterns; they were significantly aggregated only at scales of 2–4 and 2–5 m, respectively, and had random distributions at other scales. The exceptions to these findings were *S. linearistipularis* and *R. diacanthum*, which were randomly distributed at all scales. These results indicate that spatial aggregation at specific scales was common among most shrub species.

3.4. Spatial Interactions of Woody Plants—Competition between Shrubs and Adult Trees and Mutualism between Dominant Shrub Species and Juvenile Trees

As shown in Figure 5, there was a negative relationship between the presence of old trees and most shrub species. For example, there was significant negative relationship between old trees and some shrubs, e.g., *B. fruticosa* and *S. aquilegifolia*, at all scales (0–50 m). However, other shrubs had significant negative relationships with old trees only at certain scales—for example, *S. myrtilloides*, at the 10–50 m scales; *S. linearistipularis*, at the 10–20 m scales; *S. gordejevii*, at the 0–30 m scales; and *C. microphylla*, at the 0–25 m scales. In addition, there was a positive relationship between *R. diacanthum* and old trees at the 3–6 m scales. These results showed that there was a competitive relationship between most shrubs and the old trees.

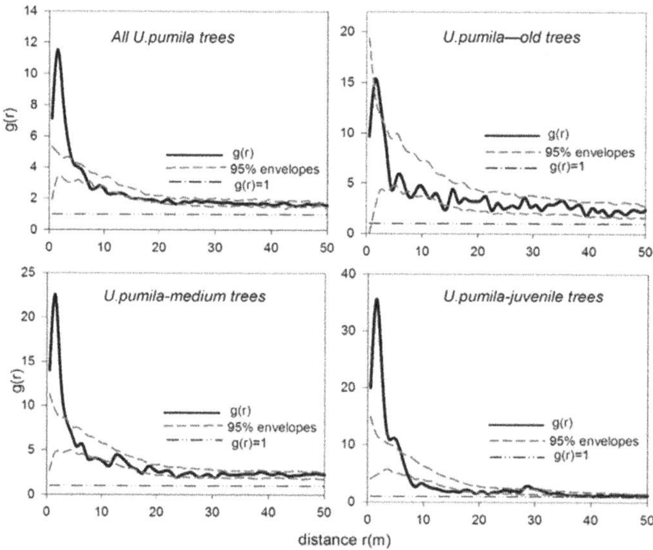

Figure 2. Univariate spatial patterns of the three *U. pumila* tree categories with the null model.

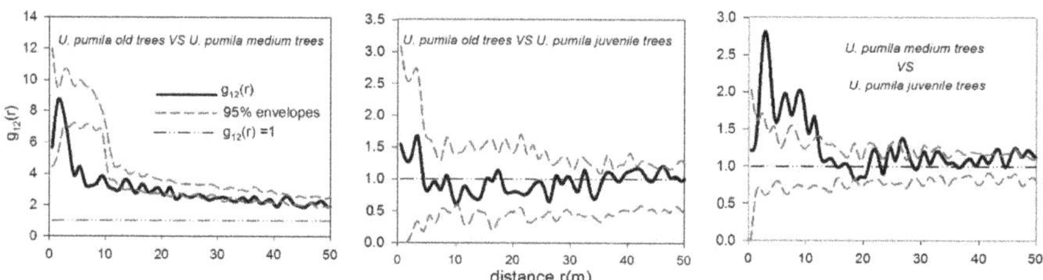

Figure 3. Bivariate spatial association between three *U. pumila* tree categories with both the toroidal shift and antecedent condition null models.

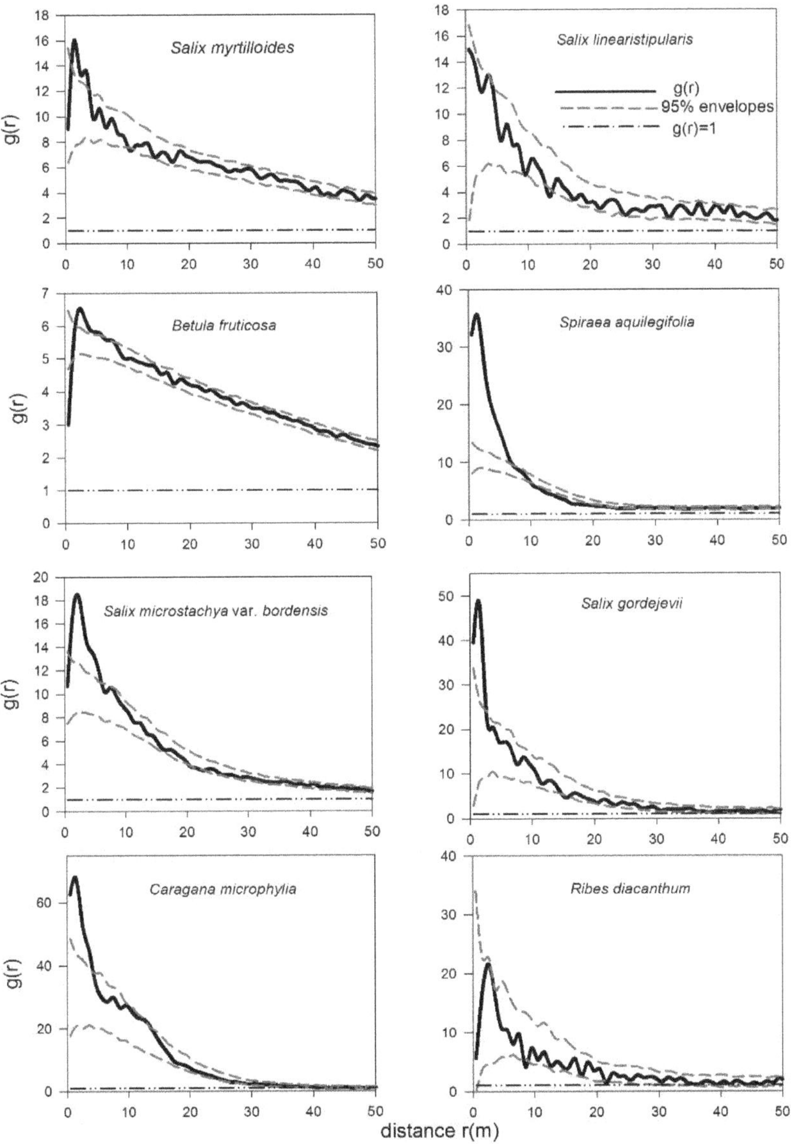

Figure 4. Univariate spatial patterns of shrub categories with the null model of complete spatial randomness.

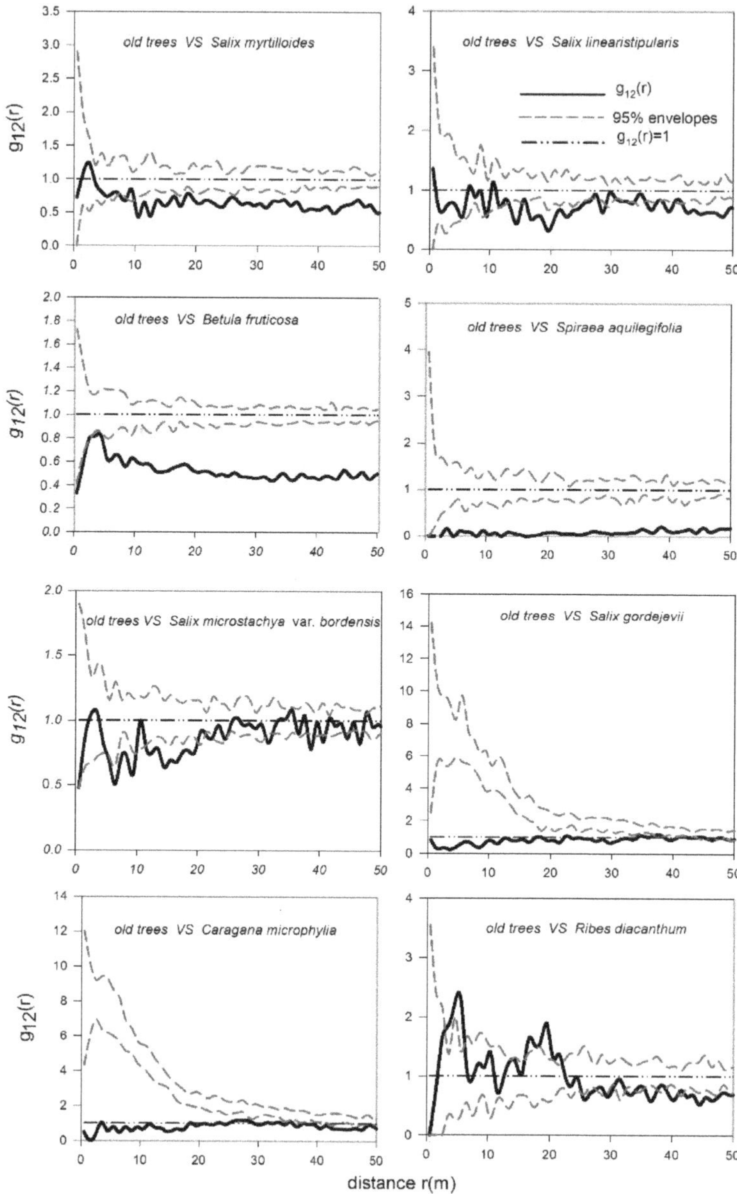

Figure 5. Bivariate spatial relationship between old *U. pumila* trees and shrubs with the independent null model.

There were negative relationships between the medium trees and most shrub species (Figure 6). For example, the presence of the shrubs *S. gordejevii*, *B. fruticosa*, *S. aquilegifolia*, and *S. microstachya* var. *bordensis* was significantly negatively associated with medium trees at all scales (0–50 m). In addition, *S. linearistipularis* was significantly negatively correlated with medium trees only at the 15–50 m scales, and no obvious relationships were observed at other scales. For *C. microphylla*, *S. gordejevii*, and *R. diacanthum*, no obvious relationships with medium trees were observed. These results also showed that there was a competitive relationship between most shrubs and medium trees.

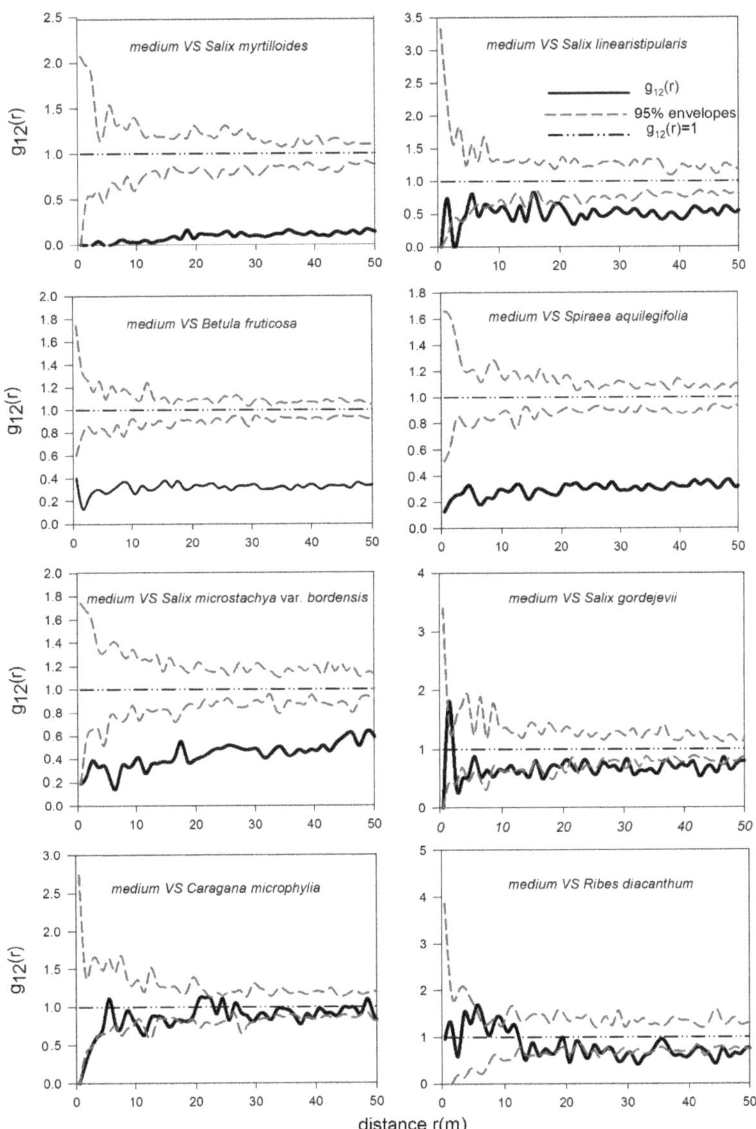

Figure 6. Bivariate spatial relationship between medium *U. pumila* trees and shrubs with the independent null model.

However, there were positive relationships between juvenile trees and some shrub species (Figure 7). *S. myrtilloides, S. aquilegifolia, S. microstachya* var. *bordensis*, and *C. microphylla* showed significant positive interactions with juvenile trees at small scales (less than 10 m). However, no other obvious relationships were observed between juvenile trees and the other shrub species.

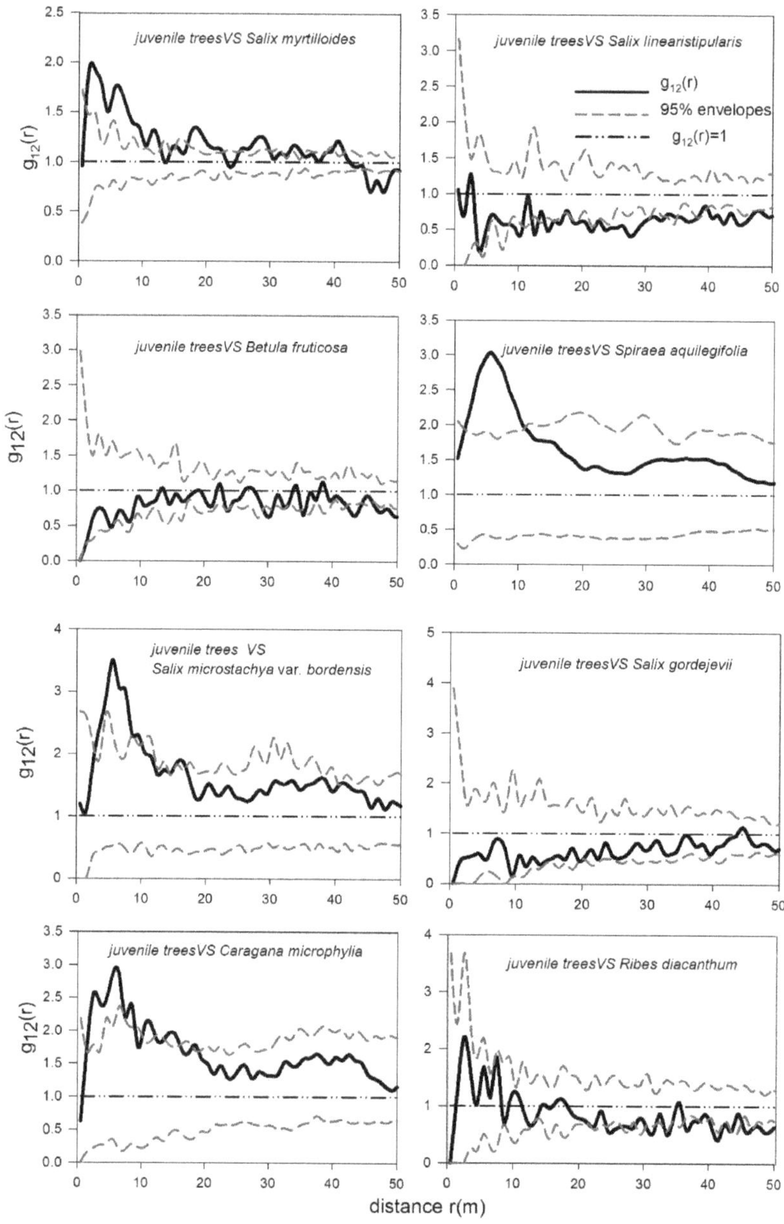

Figure 7. Bivariate spatial relationship between juvenile *U. pumila* trees and shrubs with the independent null model.

4. Discussion

4.1. Spatial Patterns of U. pumila Trees and Their Mechanism of Formation in the Temperate Savanna -Like Ecosystem

From the viewpoint of ecological niche differentiation, the large crown breadth of adult *U. pumila* trees indicates a strong demand for a broad niche [46], and its random distribution indicates strong competition among *U. pumila* trees. *U. pumila* trees of the temperate savanna normally have a larger crown width (6–12 m) and deeper roots (10–20 m depth) than *U. pumila* trees in temperate forests (4–8 m crown width, 3–5 m root depth) [42]. In this study, the old *U. pumila* trees were randomly distributed at all scales. This indicates that *U. pumila* exhibits strong self-thinning due to the intense competition among individual trees [47]. In temperate forests, self-thinning normally occurs due to light competition [48,49]. However, in this study, light was not a limiting factor, as indicated by the large tree crowns. Thus, soil moisture can be deduced to be the key limiting factor that results in trees being randomly distributed far from each other through self-thinning [50]. This deduction can also be indirectly supported by the small amount and high variability of the annual precipitation in this area, which is typically 200–350 mm [37,38]. Drought has been increasing continuously in this area since 1960, and the most severe drought in recorded history in this area occurred during 1999–2011 [51]. In addition, *U. pumila* seedlings often suffer from severe water stress during dry summers that is caused by repeated cycles of drying in the upper soil layers [52]. Wang et al. [53] noted that more than 90% of the current-year seedlings died in fenced plots because of their vulnerability to drought. Therefore, light limitation is generally replaced by increasing water stress in drought-modulated ecosystems, such as semiarid forests and temperate savanna [54].

4.2. Spatial Interactions among Trees and Shrubs—Old U. pumila Trees Inhibit the Survival of Juvenile Trees and Shrubs

In this study, a negative relationship between old *U. pumila* trees and juvenile trees was found, and significant competition between most shrub species and old *U. pumila* trees was also found. From these findings, we can conclude that the environment around old *U. pumila* trees is not suitable for the survival of juvenile trees or shrubs [53,55]. This further demonstrates that seedling regeneration in the *U. pumila*-dominated savanna-like ecosystem depends strongly on medium trees rather than on old trees. This situation is likely due to the influence of grazing. Historically, the *U. pumila*-dominated savanna-like ecosystem of the Otindag Sand Land was an important pasture source, providing a large amount of forage for livestock [51,56]. Although precipitation and soil moisture have been the main limiting factors affecting population regeneration of woody plants in arid and semiarid areas [57], grazing has become an important factor affecting population regeneration; long-term continuous heavy grazing has resulted in the destruction of topsoil and degradation of the rangeland [58]. In this savanna-like ecosystem, the area under the tree crown is usually occupied by resting livestock taking shelter from the summer heat and the intense ultraviolet radiation. Thus, the soil under the crown may be destroyed and become hardened due to the animals' trampling and reclining, and a "bare soil circle" will sometimes form under these trees [59]. This is similar to the phenomenon of piospheres caused by grazing. It was found by some studies that the soil moisture and nutrients in the piospheres are significantly lower than those in other areas; this may be the main reason for the change of vegetation composition and pattern, such as reduction in the density and production of forage, changes in the species composition of forage vegetation, shrub encroachment, and so on [59]. Some studies focus on the "bare soil circle" formed under these trees in this savanna-like ecosystem and also prove that the soil moisture and nutrients on the "bare soil circle" are lower than those outside the crown of elm [60]. These factors are unfavorable to the survival of any individual plant, including juvenile trees, shrubs, and grasses. Therefore, the disturbance of livestock caused by overgrazing may be one of the main reasons for the negative relationships between old *U. pumila* and juvenile trees and shrubs in this region.

In contrast, medium trees exhibited a significant positive association with juveniles at smaller scales. This suggests that the environmental conditions, such as light availability and soil moisture, around medium trees are more favorable for juvenile trees than those around adult trees. Some studies also show that the soil nutrients (organic matter, total nitrogen, available phosphorus) and soil moisture under the crown of medium trees are significantly higher than those outside the crown, and there are better microenvironmental conditions (light, temperature, humidity, etc.) under the crown [60]. In addition, the disturbance of livestock to medium trees is less than that to old trees.

Different shrub species showed different spatial relationships with *U. pumila* in this study. There were positive relationships between juvenile trees and some shrub species, such as *C. microphylla*. This may be because the emergence of shrubs improves soil moisture and nutrient conditions [61,62], and *C. microphylla* surely can provide some fertilization of the area due to the symbiosis with Azotobacter spp. This demonstrates that some shrubs provide some fertilization to the area in which they grow, thereby providing a favorable environment for *U. pumila* seedling growth. These results also showed that there were mutualistic relationships between juvenile trees and the shrubs *S. gordejevii* and *S. aquilegifolia*; this might be because *S. aquilegifolia* and *C. microphylla*, as spiny shrubs, can provide shelter for elm seeding, such as holding back sand and reducing damage from livestock and wind, which would protect the juvenile *U. pumila* trees [63]. Our results were similar to those of Tölgyesi et al. [9], who also found that thorny shrubs were important for supporting the biodiversity of wooded rangelands as well as facilitating the regeneration of trees by acting as nurse species.

4.3. The Development of the Original Tree-Dominated Temperate Savanna-Like Ecosystem after Shrub Encroachment

U. pumila-dominated temperate savanna-like ecosystem, once widely distributed throughout the forest-steppe ecotone on the Mongolian Plateau, is a relatively stable woody-herbaceous complex ecosystem in Northern China. Unfortunately, by the late 1990s, only a few degraded relic stands remained as scattered and fragmented patches that showed apparent regeneration failure [64]. This degradation was mainly due to irrational human land use, such as overgrazing.

Many studies clearly showed the indicative significance of shrub encroachment, which mainly depends on different factors, such as the type of ecosystem, past management practices, the current level of biodiversity, climate change, etc. [13,14]. In this region, shrubs have occupied the grassland habitat. This encroachment has led to a decrease in grass coverage and a decline in pasture quality [51]. This change is likely to become more serious with the increase in grazing pressure and the impacts of global warming. These outcomes suggest that shrub encroachment is an indicator of savanna ecosystem degradation to some extent. On the other hand, in this study, shrub encroachment was shown to promote the survival and growth of juvenile trees to some extent, which is beneficial to the reestablishment of the *U. pumila* community. However, mutualism was not observed between most shrub species and medium *U. pumila* trees in this study. This lack of mutualism may have occurred because the seedlings under the shrubs had not yet developed into medium trees at the time of this study, because the seedlings were excluded by the shrubs before developing into medium trees, or because the older seedlings may have excluded the shrubs from the habitat. If the shrubs were excluded by the older seedlings, this suggests that the emergence of shrubs in this habitat is only a phase. Conversely, if the small trees were excluded, this would be catastrophic for the ecosystem, and it would be difficult for the ecosystem to return to its original state. To determine the exact causes of these phenomena, further experimental observation is greatly needed. Regardless, this study demonstrated that shrub encroachment in this area has resulted in changes in the composition and spatial pattern of the original tree-dominated savanna.

5. Conclusions

In conclusion, shrub encroachment was observed in a temperate savanna-like ecosystem in Northern China, possibly as a result of both grazing and climate change. Shrub encroachment in this area has resulted in changes in the composition and spatial pattern of the original tree-dominated savanna. Due to severe damage to the *U. pumila*-dominated temperate savanna-like ecosystem, especially for woody vegetation, ecological degradation has occurred, such as decreasing temperate savanna-like area, loss of structural integrity, poor population regeneration, and changing spatial patterns. New management techniques and climate change mitigation practices should be implemented to replace the older, traditional management techniques and to prevent the original tree-dominated temperate savanna-like ecosystem from changing into a shrub-dominated temperate savanna-like ecosystem. A more scientific and effective management regime should be developed to maintain the stability and sustainability of this typical tree-dominated temperate savanna-like ecosystem currently and in the near future.

Author Contributions: Conceptualization, X.W., Z.S. and P.Y.; Methodology, X.Y.; Data Curation, X.W. and L.J.; Formal Analysis, X.W.; Writing—Original Draft Preparation, X.W.; Writing—Review and Editing, X.W. and P.Y; Supervision, Z.S. and Y.L. All authors have read and agreed to the published version of the manuscript.

Funding: This research was supported by the Special Foundation for National Science and Technology Basic Resource Investigation Program of China (2017FY101301), the National Key Research and Development Program of China (2016YFC0500801; 2016YFC0500804), the National Natural Science Foundation of China (31670715; 41471029; 41701249; 31200350).

Acknowledgments: We sincerely thank the local herder's family for providing their rangeland for this research. We also thank Wiegand for his generous provision of spatial analysis software.

References

1. Archer, S.R.; Andersen, E.M.; Predick, K.I.; Schwinning, S.; Woods, S.R. *Woody Plant Encroachment: Causes and Consequences*; Rangeland Systems; Springer: Cham, Switzerland, 2017; ISBN 978-3-319-46707-8.
2. Stevens, N.; Lehmann, C.E.R.; Murphy, B.P.; Durigan, G. Savanna woody encroachment is widespread across three continents. *Glob. Chang. Biol.* **2017**, *23*, 235–244. [CrossRef]
3. Naito, A.T.; Cairns, D.M. Patterns and processes of global shrub expansion. *Prog. Phys. Geogr.* **2011**, *35*, 423–442. [CrossRef]
4. Ravi, S.; D'Odorico, P.; Wang, L.; White, C.S.; Okin, G.S.; Macko, S.A.; Collins, S.L. Post-fire resource redistribution in desert grasslands: A possible negative feedback on land degradation. *Ecosystems* **2009**, *12*, 434–444. [CrossRef]
5. Ratajczak, Z.; Nippert, J.B.; Collins, S.L. Woody encroachment decreases diversity across North American grasslands and savannas. *Ecology* **2012**, *93*, 697–703. [CrossRef] [PubMed]
6. Grover, H.D.; Musick, H.B. Shrubland encroachment in southern new Mexico, USA.: An analysis of desertification processes in the american southwest. *Clim. Chang.* **1990**, *17*, 305–330. [CrossRef]
7. D'Odorico, P.; Okin, G.S.; Bestelmeyer, B.T. A synthetic review of feedbacks and drivers of shrub encroachment in arid grasslands. *Ecohydrology* **2012**, *5*, 520–530. [CrossRef]
8. Van Auken, O.W.; Bush, J. *Invasion of Woody Legumes*; Springer: Berlin/Heidelberg, Germany, 2013; p. 67.
9. Tölgyesi, C.; Bátori, Z.; Gallé, R.; Urák, I.; Hartel, T. Shrub encroachment under the trees diversifies the herb layer in a Romanian silvopastoral system. *Rangel. Ecol. Manag.* **2017**, *71*, 571–577. [CrossRef]
10. Tabares, X.; Zimmermann, H.; Dietze, E.; Ratzmann, G.; Herzschuh, U. Vegetation state changes in the course of shrub encroachment in an african savanna since about 1850ce and their potential drivers. *Ecol. Evol.* **2020**, *10*, 962–979. [CrossRef]
11. Li, Y.; Chen, J.; Cui, J.; Zhao, X.; Zhang, T. Nutrient resorption in *caragana microphylla* along a chronosequence of plantations: Implications for desertified land restoration in north china. *Ecol. Eng.* **2013**, *53*, 299–305. [CrossRef]
12. Devine, A.P.; McDonald, R.A.; Quaife, T.; Maclean, I.M.D. Determinants of woody encroachment and cover in African savannas. *Oecologia* **2017**, *183*, 939–951. [CrossRef]

13. Eldridge, D.J.; Bowker, M.A.; Maestre, F.T.; Roger, E.; Reynolds, J.F.; Whitford, W.G. Impacts of shrub encroachment on ecosystem structure and functioning: Towards a global synthesis. *Ecol. Lett.* **2011**, *14*, 709–722. [CrossRef] [PubMed]

14. Erdős, L.; Cserhalmi, D.; Bátori, Z.; Kiss, T.; Morschhauser, T.; Benyhe, B.; Dénes, A. Shrub encroachment in a wooded-steppe mosaic: Combining GIS methods with landscape historical analysis. *Appl. Ecol. Environ. Res.* **2013**, *11*, 371–384. [CrossRef]

15. Hilbig, W. *The Vegetation of Mongolia*; SPB Academic Publishing: Amsterdam, The Netherlands, 1995; 246p.

16. Fang, J.; Wang, Z.; Tang, Z. *Atlas of Woody Plants in China: Distribution and Climate*; Springer: Berlin/Heidelberg, Germany, 2011; 2020p.

17. Fu, L.; Xin, Y. Elms of China. In *The Elms: Breeding, Conservation, and Disease Management*; Dunn, C.P., Ed.; Kluwer Academic Publishers: New York, NY, USA, 2000; 384p.

18. Zhang, X.S. *Vegetation of China and Its Geographic Patterns*; Geological publishing House: Beijing, China, 2007; 1175p.

19. Zhao, H.-L.; Zhou, R.-L.; Su, Y.-Z.; Zhang, H.; Zhao, L.-Y.; Drake, S. Shrub facilitation of desert land restoration in the Horqin Sand Land of Inner Mongolia. *Ecol. Eng.* **2007**, *31*, 1–8. [CrossRef]

20. Wigley, B.J.; Bond, W.J.; Hoffman, M.T. Thicket expansion in a south african savanna under divergent land use: Local vs. global drivers? *Glob. Chang. Biol.* **2010**, *16*, 964–976. [CrossRef]

21. Venter, Z.S.; Cramer, M.D.; Hawkins, H.-J. Drivers of woody plant encroachment over africa. *Nat. Commun.* **2018**, *9*, 1–7. [CrossRef]

22. O'Connor, R.C.; Taylor, J.H.; Nippert, J.B. Browsing and fire decreases dominance of a resprouting shrub in woody encroached grassland. *Ecology* **2020**, *101*, e02935. [CrossRef]

23. Liu, F.; Wu, X.; Bai, E.; Boutton, T.W.; Archer, S.R. Spatial scaling of ecosystem c and n in a subtropical savanna landscape. *Glob. Chang. Biol.* **2010**, *16*, 2213–2223. [CrossRef]

24. Nie, W.; Yuan, Y.; Kepner, W.; Erickson, C.; Jackson, M. Hydrological impacts of mesquite encroachment in the upper san pedro watershed. *J. Arid Environ.* **2012**, *82*, 147–155. [CrossRef]

25. Li, L.; He, W.; Changcun, L.; Deli, W. Vegetation and community changes of elm (*ulmus pumila*) woodlands in northeastern china in 1983–2011. *Chin. Geogr. Sci.* **2013**, *23*, 321–330. [CrossRef]

26. Li, Z.; Wang, X.-J.; Liu, G.-H.; Niu, H.-L.; Zhen, P. Dynamics, patterns and structure of major population in ulmus pumila var. sabulosa sparse forest in hunshandake sandland. *J. Desert Res.* **2009**, *29*, 508–513. (In Chinese)

27. Chen, L.; Li, H.; Zhang, P.; Zhao, X.; Zhou, L.; Liu, T.; Hu, H.; Bai, Y.; Shen, H.; Fang, J. Shrub-encroached grassland: A new vegetation type. *Chin. J. Nat.* **2014**, *36*, 391–396. (In Chinese)

28. Su, H.; Li, Y.; Liu, W.; Xu, H.; Sun, O.J. Changes in water use with growth in *Ulmus pumila* in semiarid sandy land of northern China. *Trees* **2014**, *28*, 41–52. [CrossRef]

29. Dale, M.R.T. *Spatial Pattern Analysis in Plant. Ecology*; Cambridge University Press: Cambridge, UK, 1999; 340p.

30. Nathan, R. Long distance dispersal of plants. *Science* **2006**, *313*, 786–788. [CrossRef] [PubMed]

31. Druckenbrod, D.L.; Shugart, H.H.; Davies, I. Spatial pattern and process in forest stands with in the Virginia piedmont. *J. Veg. Sci.* **2005**, *16*, 37–48. [CrossRef]

32. Wiegand, T.; Moloney, K.A. *Handbook of Spatial Point-Pattern Analysis in Ecology*; CRC Press: Boca Raton, FL, USA, 2014; 481p.

33. He, F.; Legendre, P.; La Frankie, J.V. Distribution patterns of tree species in a Malaysian tropical rain forest. *J. Veg. Sci.* **1997**, *8*, 105–114.

34. Condit, R.; Ashton, P.S.; Baker, P.; Bunyavejchewin, S.; Gunatilleke, S.; Gunatilleke, N.; Hubbell, S.P.; Foster, R.B.; Itoh, A.; LaFrankie, J.V.; et al. Spatial patterns in the distribution of tropical tree species. *Science* **2000**, *288*, 1414–1418. [CrossRef]

35. Wang, Y.; Yang, X.; Shi, Z. The formation of the patterns of desert shrub communities on the western Ordos Plateau, China: The roles of seed dispersal and sand burial. *PLoS ONE* **2013**, *8*, e69970. [CrossRef]

36. Wu, B.; Yang, H. Spatial Patterns and Natural Recruitment of Native Shrubs in a Semi-arid Sandy Land. *PLoS ONE* **2013**, *8*, e58331. [CrossRef]

37. Chun, X.; Yong, M.; Liu, J.; Liang, W. Monitoring land cover change and its dynamic mechanism on the qehan lake basin, inner mongolia, north china, during 1977–2013. *Environ. Monit. Asses.* **2018**, *190*, 205. [CrossRef]

38. Bai, Y.F.; Wu, J.G.; Xing, Q.; Pan, Q.M.; Huang, J.H.; Yang, D.L.; Han, X.G. Primary production and rain use efficiency across a precipitation gradient on the Mongolia plateau. *Ecology* **2008**, *89*, 2140–2153. [CrossRef]
39. Editorial Committee of Flora of China. *Flora of China*, Revised ed.; Scientific Press: Beijing, China, 2013.
40. Li, S.-L.; Yu, F.-H.; Werger, M.J.A.; Dong, M.; Ramula, S.; Zuidema, P.A. Understanding the effects of a new grazing policy: The impact of seasonal grazing on shrub demography in the inner mongolian steppe. *J. Appl. Ecol.* **2013**, *50*, 1377–1386. [CrossRef]
41. Condit, R. Research in large, long-term tropical forest plots. *Trends Ecol. Evol.* **1995**, *10*, 18–22. [CrossRef]
42. Yu, P. Restoring Degraded Ecosystem in Hunshandak Sand Land through Nature Reserve. Ph.D. Thesis, Chinese Academy of Science, Beijing, China, 2005; 114p.
43. Stoyan, D.; Stoyan, H. Improving ratio estimators of second order point process characteristics. *Scand. J. Stat.* **2000**, *27*, 641–656. [CrossRef]
44. Wiegand, T.; Moloney, K.A. Rings, circles and null-models for point pattern analysis in ecology. *Oikos* **2004**, *104*, 209–229. [CrossRef]
45. Illian, J.; Penttinen, A.; Stoyan, H.; Stoyan, D. (Eds.) *Statistical Analysis and Modelling of Spatial Point Patterns*; John Wiley & Sons: Chichester, UK, 2008; 560p.
46. Kim, D.; Campbell, J.J.N. Effects of tree size, shade tolerance, and spatial pattern on the mortality of woody plants in a seminatural forest in central kentucky. *Prof. Geogr.* **2016**, *68*, 436–450. [CrossRef]
47. Skarpe, C. Spatial patterns and dynamics of woody vegetation in an arid savanna. *J. Veg. Sci.* **2010**, *2*, 565–572. [CrossRef]
48. Landry, S.; St-Laurent, M.H.; Nelson, P.R.; Pelletier, G.; Villard, M.A. Canopy cover estimation from landsat images: Understory impact on top-of-canopy reflectance in a northern hardwood forest. *Can. J. Remote Sens.* **2018**, *44*, 435–446. [CrossRef]
49. Sitch, S.; Smith, B.; Prentice, I.C.; Arneth, A.; Bondeau, A.; Cramer, W.; Kaplan, J.O.; Levis, S.; Lucht, W.; Sykes, M.T.; et al. Evaluation of ecosystem dynamics, plant geography and terrestrial carbon cycling in the lpj dynamic global vegetation model. *Glob. Chang. Biol.* **2003**, *9*, 161–185. [CrossRef]
50. Hartmann, H. Will a 385 million year-struggle for light become a struggle for water and for carbon?–How trees may cope with more frequent climate change-type drought events. *Glob. Chang. Biol.* **2011**, *17*, 642–655. [CrossRef]
51. Zhang, Z.; Zhang, B.; Zhang, X.; Yang, X.; Shi, Z.; Liu, Y. Grazing altered the pattern of woody plants and shrub encroachment in a temperate savanna ecosystem. *Int. J. Environ. Res. Public Health* **2019**, *16*, 330. [CrossRef]
52. Guo, K.; Liu, H. A comparative researches on the development of elm seedlings in four habitats in the Otindag Sandland, Inner Mongolia, China. *Acta Ecol. Sin.* **2004**, *24*, 2024–2028. (In Chinese)
53. Wang, X.; Hu, C.; Li, G.; Zuo, H. Analysis of the factors affecting seed disperal and seedling survival rate of *Ulmus pumila* in the Otindag sandy land. *Arid Zone Res.* **2011**, *28*, 542–547. (In Chinese)
54. Dai, J.; Liu, H.; Wang, Y.; Guo, Q.; Hu, T.; Quine, T.; Green, S.; Hartmann, H.; Xu, C.; Liu, X.; et al. Drought-modulated allometric patterns of trees in semi-arid forests. *Commun. Biol.* **2020**, *405*, 1–8. [CrossRef] [PubMed]
55. Wang, X.; Zhang, B.; Zhang, K.; Zhou, J.; Ahmad, B. The spatial pattern and interactions of woody plants on the temperate savanna of inner mongolia, china: The effects of alternating seasonal grazing-mowing regimes. *PLoS ONE* **2015**, *10*, e0133277. [CrossRef]
56. Qi, J.; Chen, J.; Wan, S.; Ai, L. Understanding the coupled natural and human systems in dryland east asia. *Environ. Res. Lett.* **2012**, *7*, 15202–15207. [CrossRef]
57. Priyadarshini, K.V.R.; Prins, H.H.T.; De Bie, S.; Heitkönig, I.M.; Woodborne, S.; Gort, G.; Kirkman, K.P.; Ludwig, F.; Dawson, T.E.; De Kroon, H. Seasonality of hydraulic redistribution by trees to grasses and changes in their water-source use that change tree-grass interactions. *Ecohydrology* **2015**, *9*, 217–228. [CrossRef]
58. Stumpp, M.; Wesche, K.; Retzer, V.; Miehe, G. Impact of grazing livestock and distance from water source on soil fertility in southern mongolia. *Mt. Res. Dev.* **2005**, *25*, 244–251. [CrossRef]
59. Anthony, E.; Bernard, B.; Henry, M.M.; Paul, N. Piosphere syndrome and rangeland degradation in Karamoja sub-region, Uganda. *Resour. Environ.* **2015**, *5*, 73–89. [CrossRef]
60. Zhang, Z. Spatial Distribution of Vegetation and Soil in Temperate Savanna Ecosystem, Inner Mongolia. Ph.D. Thesis, Chinese Academy Forestry, Beijing, China, 2017; 137p.

61. Dohn, J.; Augustine, D.J.; Hanan, N.P.; Ratnam, J.; Sankaran, M. Spatial vegetation patterns and neighborhood competition among woody plants in an east african savanna. *Ecology* **2016**, *98*, 478. [CrossRef]
62. Mckinley, D.C.; Blair, J.M. Woody plant encroachment by *juniperus virginiana* in a mesic native grassland promotes rapid carbon and nitrogen accrual. *Ecosystems* **2008**, *11*, 454–468. [CrossRef]
63. Schlesinger, W.H.; Raikes, J.A.; Hartley, A.E.; Cross, A.F. On the spatial pattern of soil nutrients in desert ecosystems. *Ecology* **1996**, *77*, 364–374. [CrossRef]
64. Dulamsuren, C.; Hauck, M.; Nyambayar, S. Establishment of *Ulmus pumila* seedlings on steppe slopes of the northern Mongolian mountain taiga. *Acta Oecol.* **2009**, *35*, 563–572. [CrossRef]

Publisher's Note: MDPI stays neutral with regard to jurisdictional claims in published maps and institutional affiliations.

Article

Possibilities of Speciation in the Central Sandy Steppe, Woody Steppe Area of the Carpathian Basin through the Example of *Festuca* Taxa

Károly Penksza [1], Attila Csík [2], Anna Fruzsina Filep [3,4], Dénes Saláta [1], Gergely Pápay [1,*], László Kovács [1], Kristina Varga [5], János Pauk [6], Csaba Lantos [6] and Zsuzsa Lisztes-Szabó [3]

[1] Faculty of Agricultural and Environmental Sciences, Szent István University, Páter K u 1, H-2100 Gödöllő, Hungary; penksza@gmail.com (K.P.); Salata.Denes@szie.hu (D.S.); Kovacs.Laszlo.gmbt@szie.hu (L.K.)

[2] Institute for Nuclear Research, Bem tér 18/c, H-4026 Debrecen, Hungary; csik.attila@atomki.hu

[3] Isotope Climatology and Environmental Research Centre, Institute for Nuclear Research, Bem tér 18/c, H-4026 Debrecen, Hungary; filepanna@atomki.hu (A.F.F.); lisztes-szabo.zsuzsanna@atomki.hu (Z.L.-S.)

[4] Pál Juhász-Nagy Doctoral School of Biology and Environmental Sciences, University of Debrecen, Egyetem tér 1, H-4032 Debrecen, Hungary

[5] Institutes for Agricultural Research and Educational Farm, Research Institute, University of Debrecen, Kisújszállási u. 16, H-5300 Karcag, Hungary; vargakrisztina@agr.unideb.hu

[6] Cereal Research Non-Profit Ltd., P.O. Box 391, H-6701 Szeged, Hungary; janos.pauk@gabonakutato.hu (J.P.); csaba.lantos@gabonakutato.hu (C.L.)

* Correspondence: geri.papay@gmail.com

Received: 29 October 2020; Accepted: 9 December 2020; Published: 14 December 2020

Abstract: Research Highlights: We examined the vegetation appearing in forest-steppes in the Pannon region. In the present survey taxonomical relations of the dominant *Festuca* species were examined. Background and Objectives: After deforestation and shrubcutting bare soil patches exposed to anthropogenous effects provided an opportunity for new vegetation to form. Materials and Methods: Inflorescence parameters and micromorphological characters of the leaves were examined in a new taxon and compared with two, presumably closely related, species of the genus *Festuca* L. *Festuca tomanii* Korneck & T.Gregor, with silvery leaf surface, *Festuca vaginata* W. K. and *Festuca pseudovaginata* Penksza were compared based on 24 traits of the inflorescence and their leaf anatomy studied on leaf cross-sections. Moreover, leaf micromorphological features were compared using a stereomicroscope, a scanning electron microscope completed with Energy Dispersive X-Ray Spectroscopy measurements and phytolith analysis method to establish the taxonomic applications of the micromorphological characters of the epidermis. Results: The awns of the lemma of *Festuca tomanii* were shown to be longer than those of the two other species. *Festuca vaginata* and *Festuca pseudovaginata* specimen showed low variability in inflorescence parameters but inflorescence characters were not uniform because the panicle of *Festuca tomanii* individuals was found to be bigger in the northern part than the panicles originating from the southern part of the sampled area. The phytolith assemblages of the *Festuca pseudovaginata* and *Festuca tomanii* differ from the *Festuca vaginata* in the abundance of ELONGATE SINUATE phytolith morphotype. Conclusions: we confirmed the appearance of *F. vaginata* in natural grasslands and discovered new occurrences of *F. pseudovaginata* and *F. tomanii*. *F. pseudovaginata* inhabits only the Pannon region, we found endemic and natural stands of it, but in its secondary habitats it was confirmed as a completely new species. Furthermore, taxa of disturbed vegetations are currently being examined. These habitats are potential hotspots of speciation.

Keywords: sandy grasslands; Danube-Tisza Interfluve; morphotaxonomy

---

1. Introduction

These *Festuca* taxa from the Carpathian Basin which have bowed narrow leaves were mentioned by several authors as part of *Festuca ovina* agg. [1–14]. The study in [15] arrived at the same conclusion. Those species which have continuous sclerenchyma are classified into the eu-ovina aggregate. These taxa can be identified easily based on their characteristic tissue structure and molecular genetic analyses [16–27].

The study in [28] separated a new series within the *Festuca* genus. One example is the *psammophila* series, which includes *F. polesica* Zapal, *F. vaginata* W. K., *F. psammophila* Host., F. pallens. Subsequently, [26] supplemented the series with *F. pseudovaginata* Penksza and *F. glaucina* Stohr.

In association with these groups was the taxon in question: *F. dominii* Krajina. According to [29], *F. dominii* was the dominant species in acidic sandy grasslands. *F. dominii* was described for the first time as a species by [30] and its taxonomical status was defined differently by various authors. According to [12], it was a varietas named *F. vaginata* var. *dominii*, although, [31] and [13] referred to this species as *F. vaginata* subsp. dominii, as did [2] and [11] also. Some databases do not emphasize the taxonomic importance of subspecies and *F. dominii* is considered to be a subspecies of either *F. vaginata* or *F. psammophila* as a synonym [32]. The taxonomic position of *F. dominii* Krajina was not clear, [13,14,24,25] named the taxon *F. vaginata* subsp. dominii (Krajina) P. Šmarda and [25] clarified the taxonomic status of the species and concluded that it was a subspecies of *F. psammophila* (Čelak.) Fritsch (which currently occurrs only in pine forests in Northern Europe [17]). Therefore, the accepted name of the species is *F. psammophila* subsp. *dominii* (Krajina) P. Šmarda. [33,34] examined and collected individuals belonging to *F. vaginata* in Hungarian sample sites. Based on the results, the *F. vaginata* taxon was found typically without awn. Moreover, we collected short or longer awn from the tip of the lemma, which had short awn under the tip of the lemma.

The study in [28] also distinguished the *F. trachyphylla* series, which included three species: *F. trachyphylla* (Hack.) Krajina, *F. macutrensis* Zapalowicz, *F. duvalii* (St-Yves) Storh. According to [35] *F. brevipila* (Tracey) was mistaken for decades about *F. longifolia* Thuill. In the present study it is already apparent that *F. trachyphylla* (Hack.) Kraj. and *Festuca brevipila*, although synonymous with *F. trachyphylla*, are present mainly due to coenological work conducted in this period [27,36–38]. According to [26], the *F. trachyphylla* taxon was validly referred to as *F. brevipila*. The species predominates in the pine forests of northern Europe, in sandy areas [24,25], although according to [35] it also occurs in many places in the United Kingdom. According to [39], it could be common in other sandy habitats (dunes, xerothermal sand grasslands) since it is a highly variable species with a wide ecological spectrum due to its morphological characteristics. The taxon was reported to be present in a variety of habitats but primarily in lime-poor habitats. It is common in several sources in the Koelerio-Corynephoretea association [40–42] and the Spergulo-vernalis Corynephoretea [43]. [44] Koelerion-Glaucae, [45] Sileno otitis-Festucetum, [46] described significant populations in the Potentillo-Stipetum association. The study in [47] also described it as an associative species and [48] as a character of the Violo pseudogracilis-Koelerietum splendentis ass. nov. hoc loco association. The study in [49] examined *F. ovina* agg. morphological features, including the *F. brevipila* taxon. The taxon has been extensively studied and researched. In [27] the ISSR (Inter-Simple Sequence Repeat) fingerprint analysis proved to be a useful aid in distinguishing the safe F. brevipila from other closely related species. The present study aims to describe the identifying characters. Besides the leaf schlerechyma and inflorescence features, the micromorphological characters of leaf epidermis are useful tools to identify species, especially in the family Poaceae [50–54]. Grasses deposit hydrated SiO_2 in epidermal cell walls and cells (silica bodies, phytoliths), which are genetically controlled and have taxonomical relevance [55–62]. Micromorphology, as an application, has a high priority for taxonomic studies of genus *Festuca* [63–65].

The study in [66] described a new *Festuca* taxa: *Festuca tomanii* Korneck & T.Gregor sp. nov., a fescue of sand dunes of the valleys of the northern Upper rhine, the middle main and the Bohemian Elbe. Blue green tetraploid fescues grow on base rich sands. Previously addressed as *F. duvalii*, they were described as *F. tomanii*. *F. tomanii* differs from *F. duvalii* by parchment-like sheaths of the basal leaves, of which a ± continuous sclerenchyma (*F. duvalii*: sclerenchyma mainly in three bundles). *Festuca tomanii*, a new taxon for Hungarian flora was found in the area of Újpesti Homoktövis TT [67].

Goals and hypotheses: How do the taxa found or newly discovered in identical environmental conditions differ from one another? Are there any relevant morphotaxonomical parameters the taxa can be distinguished in the field work based on? Is there a possibility for new vegetation units and species on new surfaces formed by human interventions to form, or would natural vegetation and its main species appear?

2. Materials and Methods

The specimens of the three *Festuca* taxa were transplanted in the Experimental Garden of the Genetics Institute (Szent István University, Gödöllő, Hungary) in 2018 and 2019. Inflorescences were collected from individuals grown under the same conditions for assays, leaf tissue and phytolith analysis by the dry ashing technique.

2.1. Inflorescence Measurements

For the inflorescence parameter analyses, 5-5 flowering stems were collected from each *Festuca* specimen and their parameters were measured. The origin of the individuals was as follows.

Festuca vaginata individuals. Hungary, Little Hungarian Plain (Győrszentiván (47.7167859, 17.7378554)), Danube-Tisza Interfluve (Tahitótfalu (47.799908, 19.042440)), Kisoroszi (47.808996, 18.999420), Homoktövis TT (47.602555, 19.097076), Tatárszentgyörgy (47.062065, 19.350956), Imrehegy, Inner Somogy (Böhönye (46.416657, 17.469680)), South Slovakia (near Ćenkov (47.767248, 18.527781)), Serbia (Deliblato Sands (44.955270, 21.113502)), Romania (Balta Verde (44.328763, 22.620251)). A total of 24 individuals' morphology parameters were measured as follows: 1. length of the generative stem; 2. length of inflorescence; 3. length of the longest branch on the 1st node; 4. length of the longest branch on the 2nd node; 5. length of the 4th spikelet from the top of the branch (1); 6. length of 4th spikelet from the top of inflorescence; 7. length of the 1st internode of the inflorescence. 7–15 (1): 4th spikelet from the top of the branch (2): 8. the floral number of spikelets, 9. length of upper glume, 10. length of lower glume, 11. length of the 2nd flower's lemma, 12. length of the 2nd flower's awn, 13. hair of spikelet, 14. length of the 1st flower's lemma, 15. length of the 1st flower's awn, 17–24. 4th spikelet from the top of inflorescence: 17: floral number of the spikelet, 18. length of upper glume, 19. length of lower glume, 20. length of the 2nd flower's lemma, 21. length of the 2nd flower's awn, 22. hair of spikelet 23. length of the 1st flower's lemma, 24. length of the 1st flower's awn.

Festuca pseudovaginata individuals. Hungary, Danube-Tisza Interfluve (Kisoroszi, Homoktövis TT, Szigetszentmiklós (47.650506, 19.091616), Kunpeszér, Kunadacs (47.116246, 19.259816)). A total of 17 individuals' morphology parameters were measured.

Festuca tomanii individuals. Hungary, Danube-Tisza Interfluve (Kisoroszi, Homoktövis TT, Szigetszentmiklós, Tatárszentgyörgy). A total of 20 individuals' parameters were measured.

The analyzed parameters were as follows according to [33,68].

We analyzed 24 traits for each inflorescence altogether with ordination methods and for representation discriminant analysis was used. Data were analyzed using the PAleontological STatictics Version 3.06 (PAST [69]) statistical software package. Data evaluation was performed using both classical cluster (UPGMA—Unweighted pair-group average) and ordination analysis (PCA—Principal components analysis) [70] using the former Euclidean mean distance; in the latter case, biplot and minimum spanning tree settings were used for better interpretation. For this reason, we made radar chart diagrams with polar grid typesetting of the most highlighted parameters.

2.2. Leaf Micromorphological Investigations

A total of 10 leaves of each of the three taxa were prepared for micromorphological investigations. The leaves were cleaned carefully before the examination to keep the prickles and hairs intact. Cross-sections of the leaf blades were made between the lower third and the middle of the leaves.

The air-dry cross-sections were fixed on double-sided tape mounted on an aluminum stub and were coated with gold (BIO-RAD E5000C Sputter Coater). A short section (a few mm) was cut from the same position of the leaves and fixed on an aluminum stub as described above, making the abaxial surface visible. The images were taken with a Hitachi S4300-CFE scanning electron microscope (SEM). Photos of the abaxial epidermis—and spikelets—were also taken under a Zeiss Stereo Discovery.v20 stereomicroscope. Three individual analysis measurements of the abaxial epidermis of the leaves per species were conducted using the SEM with energy dispersive X-ray fluorescence (EDX), operating at 15 kV with a detection threshold of 0.1 atom %.

Leaf blades from ten lateral shoots of the same species were collected and were treated as one single sample. Phytolith extraction was accomplished through the dry-ashing technique based on the methodological guidelines published earlier by [71–73]. The ashes were mixed thoroughly and mounted on light microscope slides in immersion oil and observed under an Alpha Euromex CMEX-5 polarized light microscope at a magnification of 400x. The phytolith samples were stored in Eppendorf tubes in the Phytolith Collection of Isotope Climatology and Environmental Research Centre (ICER) with a laboratory code of 115.2108.1-3.

A total of 3000 pieces of identifiable plant opal particles—phytoliths—per species were counted in adjacent but not in overlapping lines across the coverslip (with 22 mm length). The denomination of individual morphotypes was accomplished according to the International Code for Phytolith Nomenclature 2.0 (ICPN 2.0; [74]). Phytolith analysis was focused on the most abundant morphotypes, on grass silica short-cell phytoliths (GSSCP), long epidermal cells (ELONGATE) and long unicellular hairs or short prickles (hereunder trichomes with a common name) as these are relevant in the taxonomic aspect and can help to reveal the reason for the silver color of *F. tomanii* leaves. Hierarchical cluster analysis was undertaken (using PAST, [69]) on the phytolith frequency data.

2.3. Flow Cytometric Analyses

The ploidy level of the *Festuca* spp. was determined by flow cytometric analyses using a flow cytometer (CytoFLEX Flow Cytometer). The experiments were repeated at least three times.

The leaf samples (100 mg/plant) were collected from young leaves of plants. The samples were crashed in Eppendorf tubes containing 1ml Galbraith puffer and two stainless steel beads using TissueLyser II at 20 Hz for 3 min [75] The suspensions were purified using 20 μm sieves and 10 μL RNase solution was added to each sample for 60 min to eliminate the RNA content. DNA content was painted with 40 μL Propidium iodide (PI) solution (1 mg/mL) for 30 min, and the samples were measured by the flow cytometer. After the flow cytometric analyses, the ploidy levels of samples were determined based on histograms.

3. Results

3.1. Characters of Inflorescences

The spikelets of *F. vaginata* consisted of 4–5 flowers and those of F. tomanii comprised 5–6 flowers at (Figures 1 and 2). Most of the spikelets of *F. pseudovaginata* included five flowers with an infertile flower at the tip of the spikelets. Awns of the *F. vaginata* lemma were between 0.2–0.4 mm and those of *F. pseudovaginata* between 1–1.5 mm, and the awns of F. tomanii were longer than 2 mm.

Differences were found between the taxa based on the inflorescence data. The most important distinctive feature was the shorter inflorescence (13.47 ± 2.64 cm), the length of the longest branch on the 1st node (5.28 ± 1.78 cm), which were longer and larger (*F. pseudovaginata*: 6.97 ± 0.96 cm, *F. tomanii*: 8.83 ± 1.78 cm). In addition, the significantly shorter and smaller parameters measured included the length of the 4th spikelet from the top of the branch (*F. vaginata*: 6.20 ± 0.85 cm. *F. pseudovaginata*: 7.91 ± 0.52 cm, *F. tomanii*: 8.46 ± 0.61 cm), the length of the 4th spikelet from the top of the inflorescence (*F. vaginata*: 6.02 ± 0.77 cm. *F. pseudovaginata*: 8.69 ± 0.61 cm, *F. tomanii*: 8.49 ± 0.90 cm), the length of the upper glume (19: *F. vaginata*: 6.02 ± 0.77, 3.06 ± 0.55 mm, *F. pseudovaginata*: 8.69 ± 0.61, 3.70 ± 0.56 mm,

F. tomanii: 8.49 ± 0.90, 3.80 ± 0.37 mm), the length of the lower glume (20 *F. vaginata*: 2.10 ± 0.22, 2.31 ± 0.42 mm, *F. pseudovaginata*: 2.79 ± 0.19, 2.58 ± 0.41 mm, *F. tomanii*: 2.48 ± 0.23, 2.69 ± 0.33 mm), the length of 1st flower's awn (24: *F. vaginata*: 0.18 ± 0.16, 0.12 ± 0.11 mm, *F. pseudovaginata*: 1.64 ± 0.20, 1.51 ± 0.19 mm, *F. tomanii*: 2.13 ± 0.30, 1.93 ± 0.34 mm) and the length of 2nd flower's awn (21: *F. vaginata*: 0.17 ± 0.08, 0.21 ± 0.14 mm, *F. pseudovaginata*: 1.34 ± 0.16, 1.47 ± 0.20 mm, *F. tomanii*: 2.15 ± 0.26, 2.21 ± 0.36 mm).

Figure 1. Typical spikelets of *Festuca vaginata* W. K. (**A**), *Festuca pseudovaginata* Penksza (**B**) and *Festuca tomanii* Korneck & T.Gregor (**C**).

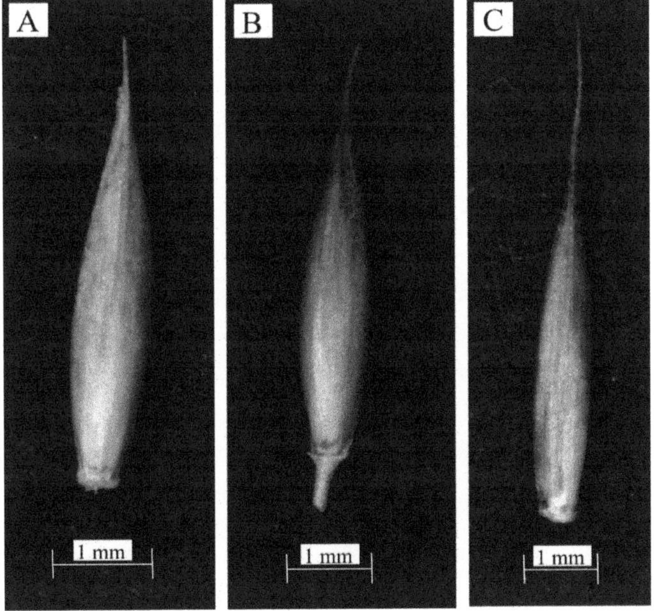

Figure 2. Tipycal lemma of *F. vaginata* (**A**), *F. pseudovaginata* (**B**) and *F. tomanii* (**C**).

The length of the generative stem showed no difference between the three taxa, and was omitted when analyzing the data. The hair of the spikelet was also not informative

The minimum spanning tree (Figure 3) and biplot options of the ordination (PCA) analysis highlighted the most responsible morphological features for species differences (Figure 4). *Festuca vaginata* separates from the other two species at a great degree. Within *F. vaginata*, there are two more groups, based on geographical location. Inflorescneces of samples originating from the southern part of the area (Balta Verde, Deliblát, Imrehegy, Tatárszentgyörgy) are relatively larger, which make them separate. The separation of *F. vaginata* was represented by the following stamps: the other two species had a much longer inflorescence (13.5 cm), the longest branch on the first node and the short awn of lemma

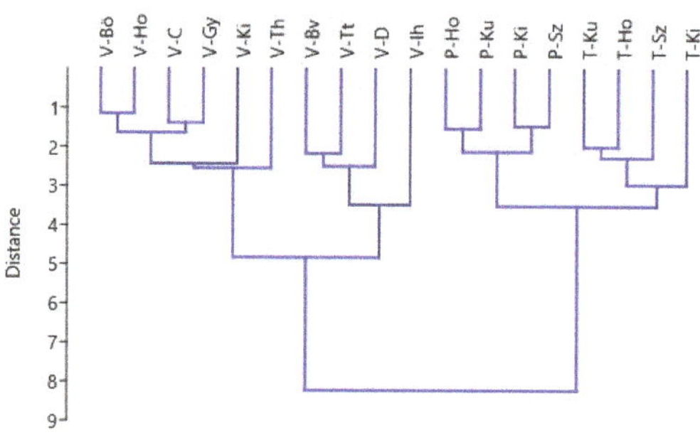

Figure 3. Classification inflorescence parameters of investigated *Festuca* taxa (V: *F. vaginata*, P: *F. pseudovaginata*, H: *F. tomanii*, Bö: Böhönye, Ho: Momoktövis TT, C: Ćenkov, Gy: Győrszentiván, Ki: Kisoroszi, Th: Tahitótfalu, Vv: Balta Verde, Tt: Tatárszentgyörgy, D. Deliblato, Ih: Imrehegy, Ku: Kunpeszér, Kunadacs, Sz: Szigetszentmiklós).

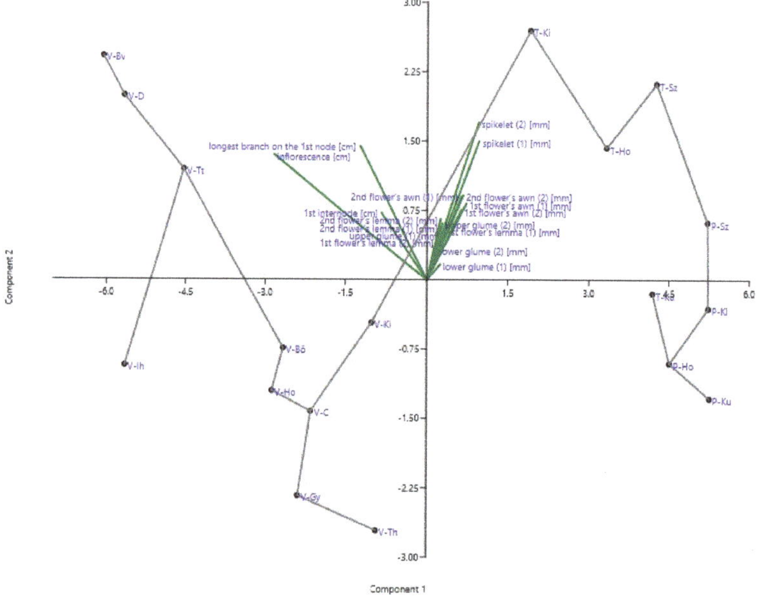

Figure 4. Classification inflorescence parameters of investigated *Festuca* taxa (V: *F. vaginata*, P: *F. pseudovaginata*, H: *F. tomanii*, Bö: Böhönye, Ho: Momoktövis TT, C: Ćenkov, Gy: Győrszentiván, Ki: Kisoroszi, Th: Tahitótfalu, Vv: Balta Verde, Tt: Tatárszentgyörgy, D. Deliblato, Ih: Imrehegy, Ku: Kunpeszér, Kunadacs, Sz: Szigetszentmiklós).

Figure 4 also highlights the morphological parameters which made *F. pseudovaginata* and *F. tomanii* different. The spikelet was also a distinctive feature of *F. vaginata* and *F. pseudovaginata* of the fourth spikelet from the top of the branch, as the fourth spikelet from the top of the branch. In the case of *F. tomanii*, this length was reversed and the fourth spikelet from the top of the branch was shorter. In these cases, similarly to *F. vaginata*, there was a difference according to their geographical distribution with the size of the specimens (Ku: Kunpeszér, Kunadacs) from the southern part of the studied areas and the south part (central Kiskunság) being smaller (north part: 8.18 ± 0.55 mm, south part: 7.56 ± 0.14 mm).

Differences between *F. pseudovaginata* and *F. tomanii*: length of the 2nd flower's awn, length of the 2nd flower's lemma, the length of the lemma awn, length of upper glume, length of 1st flower's lemma, length of 1st flower's awn.

A radar chart with polar grid type options illustrate differences in the taxa studied. The four morphological parameters were highlighted (Figure 5). The length of inflorescence (Figure 5A), which shows that the length values of *F. vaginata* were almost twice as long as the values of *F. pseudovaginata* inflorescence and the *F. tomanii* inflorescence was between the two. The difference in the length of the fourth spikelet from the top of the branch (Figure 5B) was barely significant but here the species differences could be detected, with the highest values appearing in *F. pseudovaginata* and the lowest in *F. vaginata*. The most striking differences were the length of the second flower's awn (Figure 5C). The awn lemma in *F. vaginata* was absent or very short, the longest in the *F. tomanii* taxon. The length of the upper glume (Figure 5D) had a smaller difference in size, with the highest values given by *F. pseudovaginata* and the smallest length values given by *F. vaginata*.

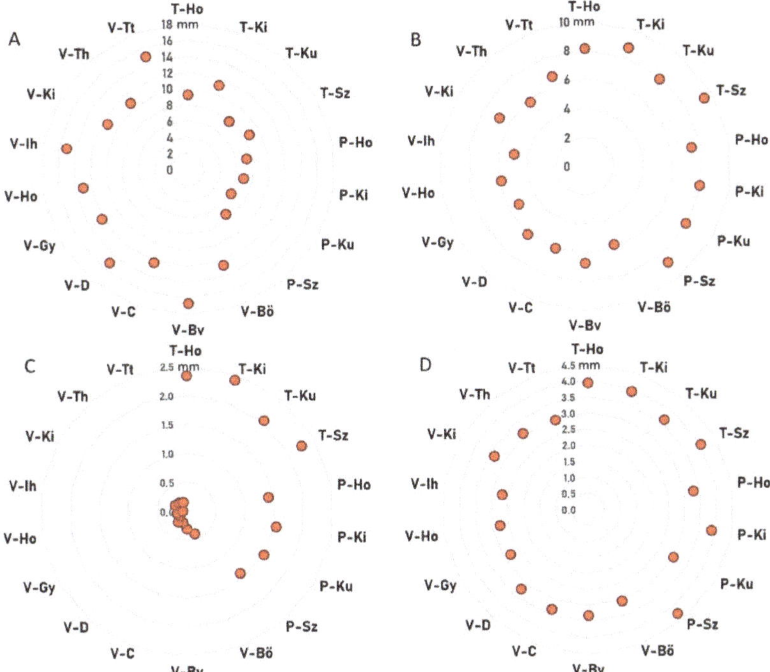

Figure 5. Radar chart with polar grid type options of some morphologycal marcs of investigated *Festuca* taxa (**A**: the length of inflorescence, **B**: length of the 4th spikelet from the top of branch, **C**: the length of the 2nd flower's awn, **D**: length of upper glume, V: *F. vaginata*, P: *F. pseudovaginata*, H: *F. tomanii*, Bö: Böhönye, Ho: Momoktövis TT, C: Ćenkov, Gy: Győrszentiván, Ki: Kisoroszi, Th: Tahitótfalu, Vv: Balta Verde, Tt: Tatárszentgyörgy, D. Deliblato, Ih: Imrehegy, Ku: Kunpeszér, Kunadacs, Sz: Szigetszentmiklós).

3.2. Leaf Micromorphology

3.2.1. Anatomy

Leaf anatomy of each of the three taxa was characterized by the number of vascular bundles of 7–9 (Figure 6). The adaxial leaf surface was covered by trichomes. The position of the sclerenchyma band was annular at the three species. The only difference we found was that one to three indenting epidermal cells interrupted the sclerenchyma ring in the leaf of *F. tomanii* at both sides near the middle vascular bundle.

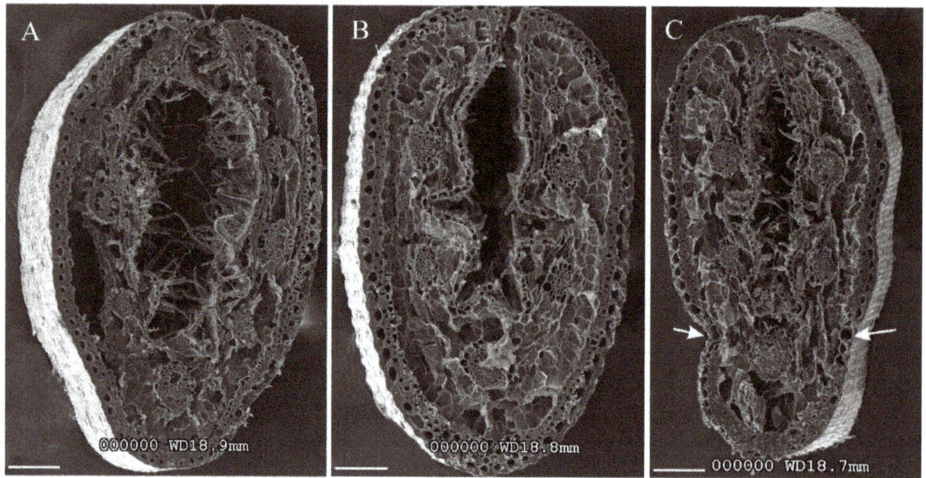

Figure 6. SEM pictures of the typical leaf cross sections. (**A**) *F. vaginata* (**B**) *F. pseudovaginata* (**C**) *F. tomanii*. White arrows show the interruptions of the continuous sclerenchyma ring at *F. tomanii*. The line represents 100 µm.

3.2.2. Phytoliths of the Leaves

Approximately 500–1000 phytolith microphotos were taken of every species with a total of 9000 (3000 per species) classified silica bodies in them (Figure 7). Several small pieces of silicified tissue were found because the cells adhered to each other sufficiently during the extraction process. Five phytolith morphotypes were counted including (following the ICPN 2.0) grass silica short-cell phytoliths (GSSCP, RONDELS), epidermal long cells with different ornaments (ELONGATE ENTIRE, ELONGATE SINUATE, ELONGATE DENTATE) and silicified trichomes (ACUTE BULBOSUS) (Table 1). There were no significant differences found in the frequency of the GSSCPs and trichomes between the species. The frequency of the ELONGATE cells was bigger at *F. pseudovaginata* and *F. tomanii* than at *F. vaginata* but the differences are not considerable. However, the distribution of the ELONGATE morphotypes represented some differences. Most of the ELONGATE phytoliths of *F. vaginata* were ELONGATE ENTIRE morphotype (81.2%) but most of the ELONGATE phytoliths of *F. pseudovaginata* and *F. tomanii* belonged to the ELONGATE SINUATE morphotype (Table 1). As a result of the hierarchical cluster analysis (Figure 8), *F. pseudovaginata* and *F. tomanii* were close to each other based on their phytolith assemblages.

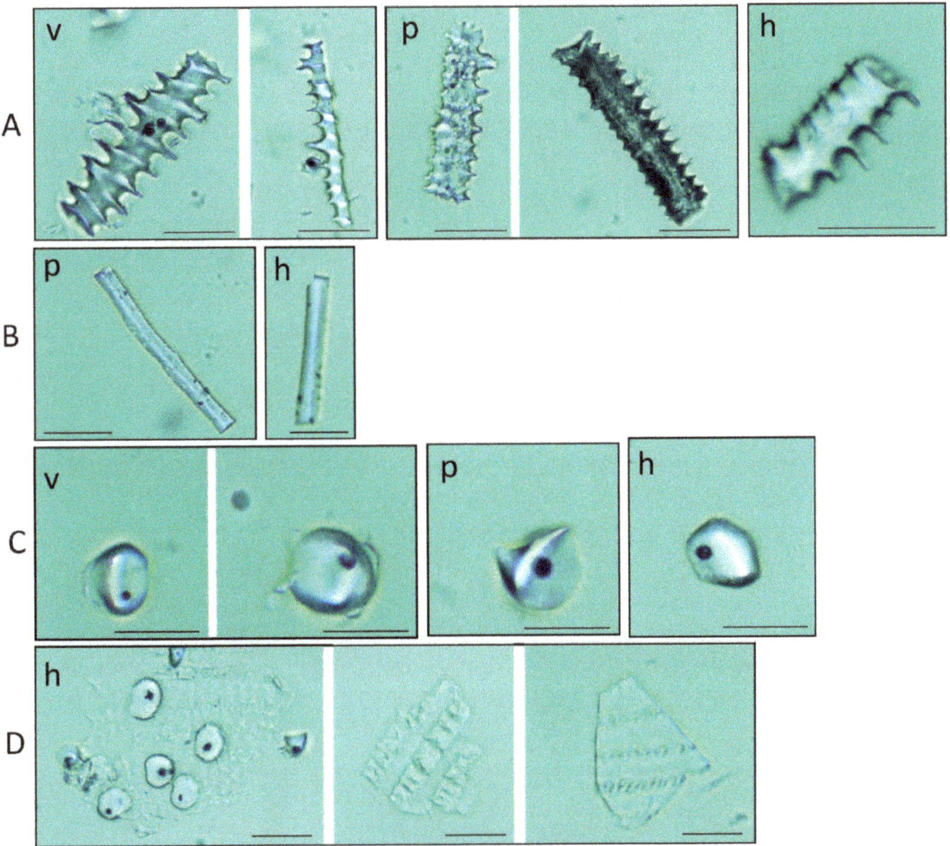

Figure 7. Light microscopic pictures of phytoliths. (**A**) ELONGATE DENTATE (**B**) ELONGATE ENTIRE (**A,B**) Silicified epidermal long cells) (**C**) RONDEL (silicified epidermal short cell), (**D**) epidermal tissue fragments with short cells and ELONGATE SINUATE phytoliths in them. (v) *F. vaginata* (p) *F. pseudovaginata* (h) *F. tomanii*. The line represents 20 μm.

Table 1. Frequency (%) of phytolith morphotypes in the leaves of *Festuca* species (sum of all the phytoliths is 100%). Frequency of the ELONGATE morphotypes means the percentage of the amount of ELONGATE cells (sum of the ELONGATE cells is 100%). GSSCP = grass silica short-cell phytolith.

	GSSCP (%)	Elongate Entire (%)	Elongate Sinuate (%)	Elongate Dentate (%)	Elongate (%)	Acute Bulbosus (%)
F. vaginata	88.3	81.2	11.9	6.9	10.7	1.0
F. pseudovaginata	81.6	39.2	52.5	8.3	17.7	0.7
F. tomanii	84.5	39.6	51.4	9.0	14.5	1.0

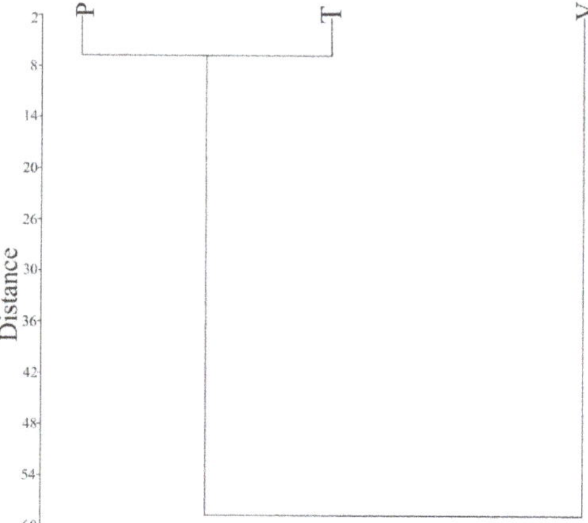

Figure 8. Hierarchical cluster dendrogram showing *Festuca* species grouping based on their phytolith assemblages (single linkage, Euclidean distance). V: *F. vaginata*, P: *F. pseudovaginata*, T: *F. tomanii*.

3.2.3. Micromorphological Characters of the Epidermis

Because *F. tomanii* leaves have silver coloration that can help to identify this taxon, differences in the micromorphological features of the abaxial leaf epidermis were expected to be found. The abaxial, dorsal epidermis of the leaves of *F. vaginata* was smooth, with short cells and the stomata submerged (Figure 9). Only a few short trichomes (20–30 µm) were found in the abaxial epidermis of the *F. pseudovaginata* leaves, sparsely at the leaf margins. However, the abaxial epidermis of the *F. tomanii* leaves had longer trichomes (with 30–100 µm) occurring more frequently (Figure 9).

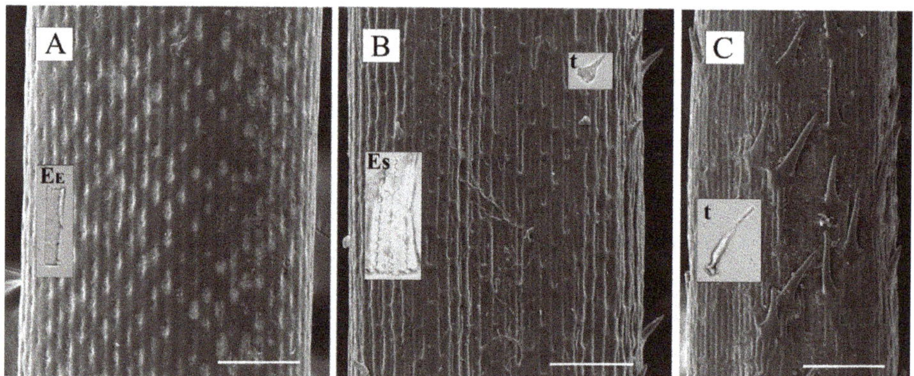

Figure 9. SEM pictures of the abaxial leaf surfaces: (**A**) *F. vaginata* (**B**) *F. pseudovaginata* (**C**) *F. tomanii*. Embedded figures are light microscopic pictures of phytoliths: EE ELONGATE ENTIRE phytoliths typical for *F. vaginata*, ES ELONGATE SINUATE phytoliths typical for *F. pseudovaginata* and *F. tomanii*, t trichomes typical for the abaxial surfaces of leaves at two latest species. The line represents 100 µm.

Moreover, silicified long cells were found in the abaxial surfaces of *F. tomanii* leaves under a stereomicroscope but there were no similar silicified cells in the epidermis of *F. vaginata* or *F. pseudovaginata* (Figure 10). EDX measurements supported the higher silicon content of *F. tomanii* leaves. The mean Si atom % values of this leaf surface were the following: at *F. vaginata* 3.65 atom %, *F. pseudovaginata* 3.50 atom % and at *F. tomanii* 14.2 atom %.

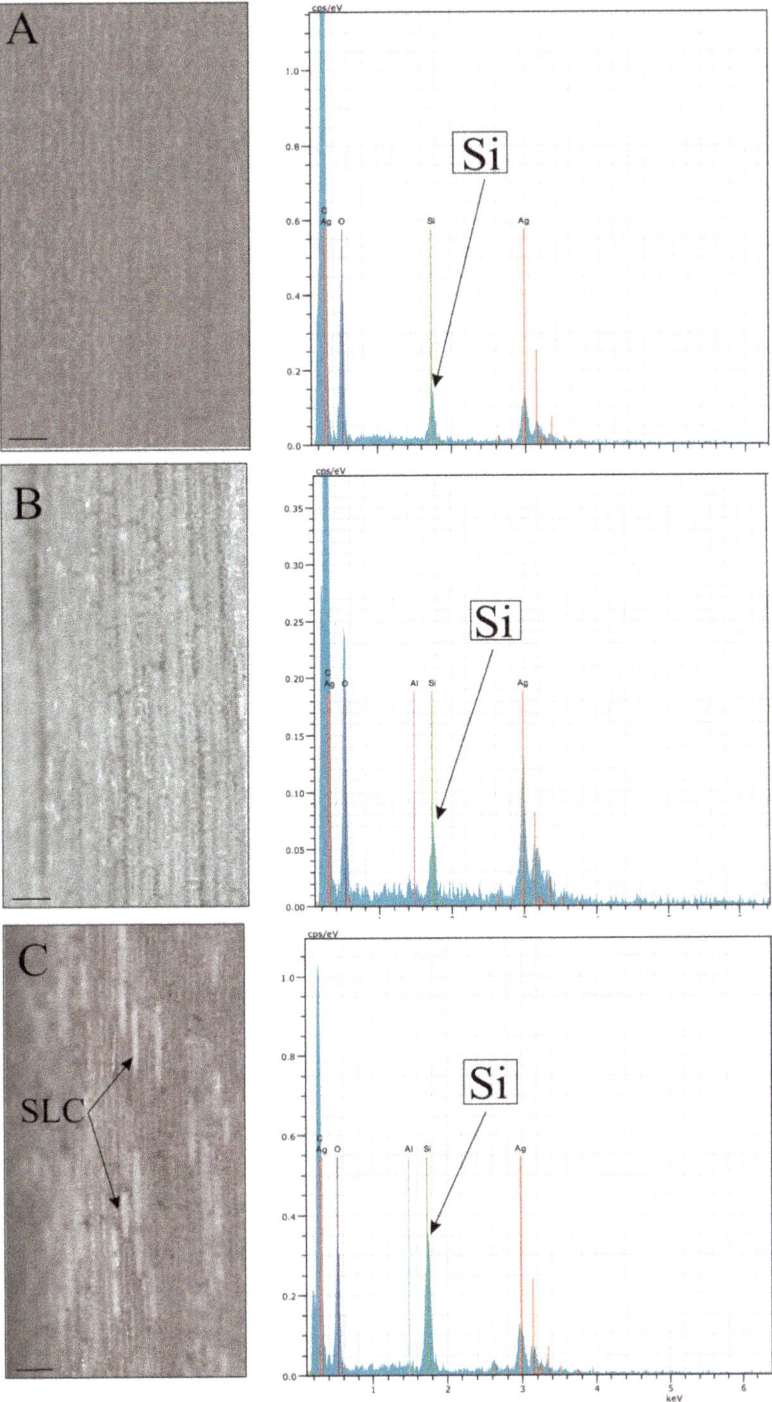

Figure 10. Stereomicroscopic pictures of the abaxial leaf surfaces: (**A**) *F. vaginata* (**B**) *F. pseudovaginata* (**C**) *F. tomanii*, with the EDX element spectrum diagrams near them. SLC: silicified long cells in the epidermis of *F. tomanii*. The line represents 50 µm.

3.3. Determination of Ploidy Level

The ploidy levels of *Festuca* spp. were determined using flow cytometric analyses. The measurements revealed the differences of DNA content among the tested samples (Figure 11). The relative DNA content was two times higher in samples of *Festuca pseudovaginata* (Figure 11B) and *Festuca tomanii* (Figure 11C) than in the sample of the diploid *Festuca vaginata* species (Figure 11A).

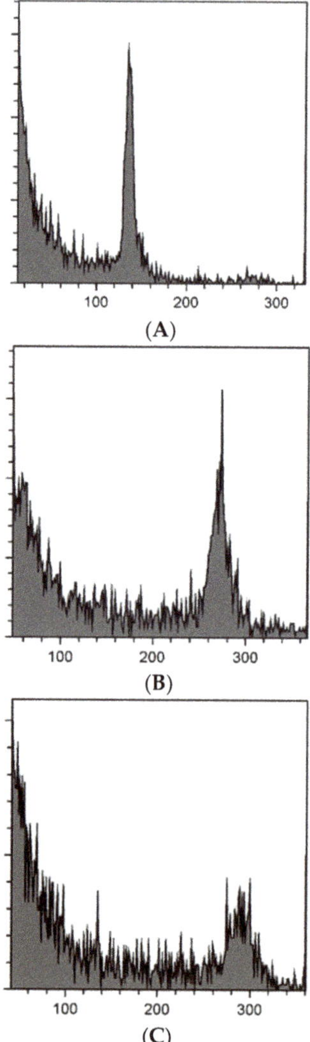

Figure 11. Flow cytometric analyses of *Festuca* spp.: histograms demonstrate the relative DNA content of (**A**) *Festuca vaginata*, (**B**) *Festuca pseudovaginata* and (**C**) *Festuca tomanii*.

4. Discussion

One of the most important identification keys to these *Festuca* taxa was the length of the awn of the lemma. Awns of *F. vaginata* were missing or very short (0.2–0.4 mm), confirming the findings of [25,26] and [34]. Awn of the lemma of *F. pseudovaginata* is longer (1.2–1.8 mm) according to [76].

Awn of the lemma of *F. tomanii* is significantly longer than 2 mm [77], which was confirmed by the present study results.

Among the inflorescence parameters there were morphological markers that were not suitable for distinguishing the three species examined, such as the length of the generative stem, floral number of spikelet, length of upper glume and hairiness. Beyond these marks, *F. vaginata* was distinguished by the other parameters studied, including the longer inflorescence branch and the significantly longer lower inflorescence branch. The spikelet was the shortest of the three taxa examined. According to the spikelet, *F. pseudovaginata* and *F. tomanii* can be distinquished. The studies in [78,79] highlighted this fact. According to [78,79], samples must be taken from one particular point of the panicle. The present data confirmed [78,79] finding that a spikelet at a given position should be examined. In the case of F. goals and no hypotheses, the fourth spikelet at the lower inflorescence branch was longer than the fourth spikelet at the apex of the inflorescence, which is also a good distinguishing morphological parameter of identification.

The distribution of the different phytolith morphotypes did not answer the question of what caused the silver coloration of the *F. tomanii* leaves. There were no more ELONGATE phytoliths or trichomes in the *F. tomanii* leaves than in the two other species but the phytolith analysis highlighted the differences between the micromorphological features of the abaxial epidermis surfaces of the studied *Festuca* leaves, namely that most of the ELONGATE phytoliths of *F. vaginata* were the ELONGATE ENTIRE morphotype but most of the ELONGATE phytoliths of *F. pseudovaginata* and *F. tomanii* belonged to the ELONGATE SINUATE morphotype. This finding supports the results of the inflorescence data analysis of these species and confirm the usefulness of the quantitative phytoliths analysis for revealing a new taxonomic character to distinguish different species of a grass genus [77]. As all the three taxa had trichomes in the adaxial surfaces of their leaves, we could not find considerable differences among the species concerning the number of the trichomes in the phytolith assemblages.

However, it was a reliable parameter in their identification key that the abaxial surfaces of the *F. tomanii* leaves bore trichomes, which may be the reason for the silvery epidermis. It is not clear why the silicified long epidermal cells observed under a stereo microscope were not represented in the phytolith assemblages of *F. tomanii* in larger numbers. On the other hand, the EDX measurements proved the high silica content of the abaxial epidermis of *F. tomanii* leaves, with more than triple Si atom % value.

Based on the length of the spikelet, the size of *F. tomanii,* the individuals in the middle of the Kiskunság were smaller, probably due to the adaptation to the drier and warmer habitat [80,81] Another strategy to adapt to it is the more intensive silica accumulation [82–85], which is also a characteristic feature of *F. tomanii*. As this taxon has special morphological and anatomical characters according to its distinct area and habitat.

Based on our results, we confirmed the appearance of *F. vaginata* in natural grasslands and discovered new occurrences of *F. pseudovaginata* and *F. tomanii*. *F. pseudovaginata* inhabits only the Pannon region; we found endemic and natural stands of it, but in its secondary habitats it was confirmed as a completely new species. Furthermore, taxa of disturbed vegetations are currently being examined. These habitats are potential hotspots of speciation.

On bare soil surfaces of areas exposed to anthropogenous effects, two species of the genus *Festuca* became dominant.

5. Conclusions

We examined the species pool of the sandy grasslands in the steppe, forest-steppe zone of the central region of the Carpathian Basin, with greater consideration of the dominant *Festuca* species. Three of them occurred in the open sandy grasslands: *F. vaginata*, *F. pseudovaginata* (the latter appeared on disturbed grasslands after deforestation) and *F. tomanii* as a new species in the Carpathian Basin. We compared the morphotaxonomy of these taxa, especially the characteristics which can be useful when distinguishing them during field work. Micromorphological examination of the epidermis and phytolith analysis were new elements. These can refine the on-the-spot identification of these taxa.

Author Contributions: Conceptualization—K.P.; Data Curation—A.C.; Investigation—A.F.F., D.S., G.P., L.K., K.V., J.P., C.L., Z.L.-S. All authors have read and agreed to the published version of the manuscript.

Funding: The work was funded by OTKA K-125423, EFOP-3.6.2-16-201700001 and TKP2020-NKA-21.

Acknowledgments: The work was supported by OTKA K-125423, TKP2020-NKA-21 and EFOP-3.6.2-16-201700001. We acknowledge the general support of the Kiskunság National Park, the Duna-Ipoly National Park, the Fertő-Hanság National Park, Budapest Waterworks, The Mayor's Office, Budapest. Author Z.L.-S. was supported by the European Union and the State of Hungary, co-financed by the European Regional Development Fund in project GINOP-2.3.2.-15-2016-00009 'ICER.'

Conflicts of Interest: The authors declare no conflict of interest.

References

1. Adler, W.; Oswald, K.; Fischer, R. *Exkursionflora von Österreich*; Excursion Flora of Austria: Ulmer, Austria, 1994.
2. Dostal, J. *Nová Květena ČSSR [New Flora of CSSR] (Vol. 2)*; Academia Praha: Praha, Czech Republic, 1989.
3. Király, G. *Új Magyar Füvészkönyv. Magyarország Hajtásos Növényei. Határozókulcsok*; Aggteleki Nemzeti Park Igazgatóság: Tengerszem, Hungary, 2009.
4. Horánszky, A.; Jankó, B.; Vida, G. Zur Biosystematik der *Festuca ovina*-gruppe in Ungarn [Biosystematics of *Festuca ovina*-group in Hungary]. *Ann. Univ. Sci. Bp. Sect. Biol.* **1971**, *13*, 95–101.
5. Májovszký, J. Adnotationes ad species gen. *Festuca* florae Slovakiae additamentum I. *Acta Fac. Rer. Nat. Univ. Comen.* **1962**, *7*, 317–355.
6. Nyárády, E.I.; Nyárády, A. Studie über die Arten der Sektion Ovinae Fr. der Gattung *Festuca* in der RVR [Study on species of *Festuca ovina* group in RVR]. *Revue Roum. Biol. Sér. Botan.* **1964**, *9*, 99–172.
7. Patzke, E. Vorschlag zur Gliederung der *Festuca ovina* L.-Gruppe in Mitteleuropa [Recommendation for classification of the *Festuca ovina* L. group in Central-Europe]. *Österreichische Bot. Z.* **1961**, *108*, 505–507. [CrossRef]
8. Patzke, E. Zur Kenntnis der Sammelart *Festuca ovina* L. im südlichen Niedersaschen [Study on *Festuca ovina* L. in Lower Saxony]. *Götting. Flor. Rundbr.* **1968**, *4*, 14–17.
9. Pils, G. Systematik, Karyologie und Verbreitung der *Festuca* uaJesiaca-Gruppe (Poaceae) in Österreich und Südtirol. *Syst. Karyology Distrib. Festuca Val. Group Austria South Tyrol* **1984**, *24*, 35–77.
10. Săvulescu, T. *Flora Republicii Socialiste România*; Editura Academiei Republicii Socialiste România: Bucureşti, Romania, 1972.
11. Schwarzová, T. Beitrag zur Lösung taxonomischer Probleme der *Festuca vaginata* W. et K. und F. psammophila Hack. *Acta Fac. Rer. Nat. Univ. Comen. Bot.* **1967**, *14*, 381–414.
12. Soó, R. Festuca Studien. *Acta Bot. Acad. Sci. Hung.* **1955**, *2*, 187–221.
13. Soó, R. Zeitgemässe Taxonomie der *Festuca ovina*-gruppe [Recent taxonomy of *Festuca ovina*-group]. *Acta Bot. Sci. Hung.* **1973**, *18*, 363–377.
14. Soó, R. *A Magyar Flóra és Vegetáció Rendszertani-Növényföldrajzi Kézikönyve. II [Taxonomical and Phytogeographical Manual of the Hungarian Flora and Vegetation]*; Akadémiai Kiadó: Budapest, Hungary, 1973.
15. Hackel, E. Monographia Festucarum europaearum. Auctore, Eduardo Hackel. In *Monographia Festucarum europaearum. Auctore, Eduardo Hackel.*; Smithsonian Institution: Washington, DC, USA, 1882.
16. Isobe, S.N. Sustainable use of genetic diversity in forage and turf breeding. *Ann. Bot.* **2013**, *111*, 41–45. [CrossRef]
17. Bednarska, I.; Kostikov, I.; Tarieiev, A.; Stukonis, V. Morphological, Karyological and Molecular Characteristics of *Festuca* arietina Klok.—A Neglected Psammophilous Species of the *Festuca valesiaca* agg. from Eastern Europe. *Acta Biol. Crac. Bot.* **2017**, *59*, 35–53. [CrossRef]
18. Galli, Z.; Penksza, K.; Kiss, E.; Bucherna, N.; Heszky, L. *Festuca* fajok molekuláris taxonómiai vizsgálata a *F. ovina* csoport RAPD és AP-PCR analízise [Molecular taxonomic analysis of *Festuca* species: RAPD and AP-PCR analysis of the *F-ovina* group]. *Növénytermelés* **2001**, *50*, 375–384.
19. Galli, Z.; Penksza, K.; Kiss, E.; Sági, L.; Heszky, L.E. Low variability of internal transcribed spacer rDNA and trnL (UAA) intron sequences of several taxa in the *Festuca ovina* aggregate (Poaceae). *Acta Biol. Hung.* **2006**, *57*, 57–69. [CrossRef] [PubMed]

20. Gaut, B.S.; Tredway, L.P.; Kubík, C.; Gaut, R.L.; Meyer, W. Phylogenetic relationships and genetic diversity among members of the *Festuca*-Lolium complex (Poaceae) based on ITS sequence data. *Plant Syst. Evol.* **2000**, *224*, 33–53. [CrossRef]

21. Loureiro, J.; Kopecký, D.; Castro, S.; Santos, C.; Silveira, P. Flow cytometric and cytogenetic analyses of Iberian Peninsula *Festuca* spp. *Plant Syst. Evol.* **2007**, *269*, 89–105. [CrossRef]

22. Šmarda, P. DNA ploidy levels and intraspecific DNA content variability in Romanian fescues (*Festuca*, *Poaceae*) measured in fresh and herbarium material. *Folia Geobot. Phytotaxon.* **2006**, *41*, 417–432. [CrossRef]

23. Šmarda, P. DNA ploidy level variability of some fescues (*Festuca subg. Festuca, Poaceae*) from Central and Southern Europe measured in fresh plants and herbarium specimens. *Biology* **2008**, *63*, 349–367. [CrossRef]

24. Šmarda, P.; Kočí, K. Chromosome number variability in central european members of the *Festuca ovina* and *F. pallens* groups (sect. *Festuca*). *Folia Geobot. Phytotaxon.* **2003**, *38*, 65–95. [CrossRef]

25. Šmarda, P.; Šmerda, J.; Knoll, A.; Bureš, P.; Danihelka, J. Revision of Central European taxa of *Festuca* ser. Psammophilae Pawlus: Morphometrical, karyological and AFLP analysis. *Plant Syst. Evol.* **2007**, *266*, 197–232. [CrossRef]

26. Šmarda, P.; Bureš, P.; Horová, L.; Foggi, B.; Rossi, G. Genome Size and GC Content Evolution of *Festuca*: Ancestral Expansion and Subsequent Reduction. *Ann. Bot.* **2008**, *101*, 421–433. [CrossRef]

27. Stukonis, V.; Armonienė, R.; Lemežienė, N.; Kemešytė, V.; Statkevičiūtė, G. Identification of fine-leaved species of genus *Festuca* by molecular methods. *Pak. J. Bot.* **2015**, *47*, 1137–1142.

28. Pawlus, M. Systematyka i rozmieszczenie gatunków grupy *Festuca ovina* L. w Polsce [Classification and distribution of *Festuca ovina* L. in Poland]. *Fragm. Florist. Geobot.* **1985**, *29*, 219–295.

29. Borhidi, A.; Kevey, B.; Lendvai, G.; Seregélyes, T. *Plant Communities of Hungary*; Akadémiai Kiadó: Budapest, Hungary, 2012.

30. Krajina, V. Adnotationes ad species generis *Festuca* in Flora Cechoslovenika exsiccata. *Acta Bot. Bohem.* **1930**, *9*, 186–220.

31. Soó, R.; Jávorka, S. A magyar növényvilág kézikönyve. In *Handbook of Hungarian Flora*; Akadémiai Kiadó: Budapest, Hungary, 1951.

32. The Plant List. Available online: www.theplantlist.org (accessed on 8 September 2020).

33. Penksza, K.; Szabó, G.; Zimmermann, Z.; Lisztes-Szabó, Z.; Pápay, G.; Járdi, I.; Fűrész, A.; S-Falusi, E. A *Festuca vaginata* alakkör taxonómiai problematikája és ennek cönoszisztematikai vonatkozásai [The taxonomic problems of the *Festuca vaginata* agg. and their coenosystematic aspects]. *Georg. Agric.* **2019**, *23*, 63–76.

34. Penksza, K. Kiegészítések a hazai *Festuca* taxonok ismeretéhez I.: A *Festuca psammophila* series *Festuca vaginata* alakköre. *Bot. Közlemények* **2019**, *106*, 65–70. [CrossRef]

35. Wilkinson, M.J. Stace, The taxonomic relationship and typification of *Festuca* brevipila Tracey and *Festuca* lemanii Bastard (Poaceae). *Watsonia* **1989**, *17*, 289–299.

36. Budak, H.; Shearman, R.; Gaussoin, R.; Dweikat, I. Application of Sequence-related Amplified Polymorphism Markers for Characterization of Turfgrass Species. *HortScience* **2004**, *39*, 955–958. [CrossRef]

37. Lonati, M.; Lonati, S. Le praterie xerofile a *Festuca* trachyphylla (Hackel) Krajina della bassa Valsesia (Piemonte, Italia) [The xerophilous *Festuca* trachyphylla (Hackel) Krajina grasslands in the lower Valsesia (Piedmont, Italy)]. *Fitosociologia* **2007**, *44*, 109–118.

38. Stukonis, V.; Lemežienė, N.; Kanapeckas, J. Suitability of narow-leaved *Festuca* species for turf. *Agron. Res.* **2010**, *8*, 729–734.

39. Dąbrowska, A. Morpho-anatomical structure of the leaves of *Festuca* trahyphylla (Hack.) Krajina in the ecological aspect. *Mod. Phytomorphol.* **2012**, *1*, 19–22. [CrossRef]

40. Gugnacka-Fiedor, W.; Adamska, E. The preservation state of the flora and vegetation of the artillery range near the city of Toruń. *Ecol. Quest.* **2010**, *12*, 75–86. [CrossRef]

41. Löbel, S.; Dengler, J. Dry grassland communities on southern Öland: Phytosociology, ecology, and diversity. *Acta Phytogeogr. Suec.* **2008**, *88*, 13–31. [CrossRef]

42. Rūsiņa, S. Nelku aira *Aira Caryophyllea* L. Latvija [*Aira caryophyllea* L. in Latvia]. *Latv. Veģetācija* **2003**, *7*, 33–43.

43. Kaczmarek, Z.; Gajewski, P.; Mocek, A.; Grzelak, M.; Knioła, A.; Glina, B. Geobotanical conditions of ecological grasslands on light river alluvial soils. *J. Res. Appl. Agric. Eng.* **2015**, *60*, 131–135.

44. Böhnert, W.; Reichhoff, L. Die Pflanzengesellschaften des Naturschutzgebietes "Steckby-Lödderitzer Forst". *Plant Communities Steckby-Lödderitzer Forst Nat. Reserve* **1978**, *15*, 106–114.

45. Fischer, W.; Kummer, V.; Pötsch, J. Zur Vegetation des Feuchtgebietes Internationaler Bedeutung (FIB) untere Havel [Vegetation of lower Havel, an internationally important wetland area]. *Nat. Landsch. Brandenbg.* **1995**, *4*, 12–18.

46. Nienartowicz, A.; Kaminski, D.; Kunz, M.; Deptuła, M.; Adamska, E. Changes in the plant cover of the dune hill in Folusz near Szubin (NW Poland) between 1959 and 2012: The problem of preservation of xerothermic grasslands in the agricultural landscape. *Ecol. Quest.* **2015**, *20*, 23. [CrossRef]

47. Kovář, P. Contribution to the syntaxonomy of the *Festuca* trachyphylla-grass-lands. *Preslia* **1980**, *52*, 217–226.

48. Di Pietro, R. New Dry Grassland Associations from the Ausoni-Aurunci Mountains (Central Italy)-Syntaxonomical Updating and Discussion on the Higher Rank Syntaxa. *Hacquetia* **2011**, *10*, 183–231. [CrossRef]

49. Stace, C.A.; Al-Bermani, A.-K.K.A.; Wilkinson, M.J. The distinction between the *Festuca ovina* L. and *Festuca rubra* L. aggregates in the British Isles. *Watsonia* **1992**, *19*, 107–112.

50. Blackman, E. Opaline silica bodies in the range grasses of southern Alberta. *Can. J. Bot.* **1971**, *49*, 769–781. [CrossRef]

51. Brown, W.V. Leaf Anatomy in Grass Systematics. *Int. J. Plant Sci.* **1958**, *119*, 170–178. [CrossRef]

52. Metcalfe, C.R. *Anatomy of the Monocotyledons: I. Gramineae*; Clarendon Press: Oxford, UK, 1960.

53. Palmer, P.G.; Tucker, A.E. A Scanning Electron Microscope Survey of the Epidermis of East African Grasses, I. *Smithson. Contrib. Bot.* **1981**, *49*, 1–84. [CrossRef]

54. Tateoka, T.; Inoue, S.; Kawano, S. Notes on Some Grasses. IX. Systematic Significance of Bicellular Microhairs of Leaf Epidermis. *Int. J. Plant Sci.* **1959**, *121*, 80–91. [CrossRef]

55. Brown, D.A. Prospects and limits of a phytolith key for grasses in the central United States. *J. Archaeol. Sci.* **1984**, *11*, 345–368. [CrossRef]

56. Fredlund, G.G.; Tieszen, L.T. Modern Phytolith Assemblages from the North American Great Plains. *J. Biogeogr.* **1994**, *21*, 321. [CrossRef]

57. Krishnan, S.; Samson, N.P.; Ravichandran, P.; Narasimhan, D.; Dayanandan, P. Phytoliths of Indian grasses and their potential use in identification. *Bot. J. Linn. Soc.* **2000**, *132*, 241–252. [CrossRef]

58. Lu, H.; Liu, K.-B. Morphological variations of lobate phytoliths from grasses in China and the south-eastern United States. *Divers. Distrib.* **2002**, *9*, 73–87. [CrossRef]

59. Mulholland, S.C. Phytolith shape frequencies in North Dakota grasses: A comparison to general patterns. *J. Archaeol. Sci.* **1989**, *16*, 489–511. [CrossRef]

60. Mulholland, S.C.; Rapp, G. A Morphological Classification of Grass Silica-Bodies. In *Phytolith Systematics*; Springer Science and Business Media LLC: Berlin, Germany, 1992; pp. 65–89.

61. Prychid, C.J.; Rudall, P.J.; Gregory, M. Systematics and Biology of Silica Bodies in Monocotyledons. *Bot. Rev.* **2003**, *69*, 377–440. [CrossRef]

62. Twiss, P.C.; Suess, E.; Smith, R.M. Morphological Classification of Grass Phytoliths. *Soil Sci. Soc. Am. J.* **1969**, *33*, 109–115. [CrossRef]

63. Pepi, M.F.; Zucol, A.F.; Arriaga, M. Comparative phytolith analysis of *Festuca* (*Pooideae: Poaceae*) species native to Tierra del Fuego, Argentina. *Botany* **2012**, *90*, 1113–1124. [CrossRef]

64. Ortúñez, E.; De La Fuente, V. Epidermal micromorphology of the genus *Festuca* L. (*Poaceae*) in the Iberian Peninsula. *Plant Syst. Evol.* **2010**, *284*, 201–218. [CrossRef]

65. Zarinkamar, F.; Jouyandeh, N.E. Foliar anatomy and micromorphology of *Festuca* L. and its taxonomic applications. *Taxon. Biosyst.* **2011**, *8*, 55–63.

66. Korneck, D.; Gregor, T. *Festuca tomanii* sp. nov., ein Dünen-Schwingel des nördlichen oberrhein-, des mittleren main- und des böhmischen Elbetales. *Kochia* **2015**, *9*, 37–58.

67. Bajor, Z. Effect of conservation management practices on sand grassland vegetation in budapest, hungary. *Appl. Ecol. Environ. Res.* **2016**, *14*, 233–247. [CrossRef]

68. Csány-Kovács, C.; Horánszky, A. Charakterisierung der *Festuca*-Populationen aufgrund der Merkmale der Rispe. *Ann. Univ. Sci. Bp. Rolando Eötvös Nomin. Sect. Biol.* **1973**, *15*, 59–74.

69. Hammer, O.; Harper, D.; Ryan, P. PAST: Paleontological Statistics Software Package for Education and Data Analysis. *Palaeontol. Electron.* **2001**, *4*, 1–9.

70. Harper, D.A.T. *Numerical Palaeobiology: Computer-Based Modelling and Analysis of Fossils and Their Distributions*; John Wiley & Sons: Hoboken, NJ, USA, 1999.

71. Weiner, S.; Albert, R. Study of Phytoliths in Prehistoric Ash Layers from Kebara and Tabun Caves Using a Quantitative Approach. In *Phytoliths-Applications in Earth Science and Human History*; Informa UK Limited: Colchester, UK, 2001; pp. 251–266.

72. Mercader, J.; Bennett, T.; Esselmont, C.; Simpson, S.; Walde, D. Phytoliths in woody plants from the Miombo woodlands of Mozambique. *Ann. Bot.* **2009**, *104*, 91–113. [CrossRef]

73. Mercader, J.; Astudillo, F.; Barkworth, M.; Bennett, T.; Esselmont, C.; Kinyanjui, R.; Grossman, D.L.; Simpson, S.; Walde, D. Poaceae phytoliths from the Niassa Rift, Mozambique. *J. Archaeol. Sci.* **2010**, *37*, 1953–1967. [CrossRef]

74. Neumann, K.; Strömberg, C.A.E.; Ball, T.; Albert, R.M.; Vrydaghs, L.; Cummings, L.S.; International Committee for Phytolith Taxonomy (ICPT). International Code for Phytolith Nomenclature (ICPN) 2.0. *Ann. Bot.* **2019**, *124*, 189–199. [CrossRef]

75. Galbraith, D.W.; Harkins, K.R.; Maddox, J.M.; Ayres, N.M.; Sharma, D.P.; Firoozabady, E. Rapid flow cytometric analysis of the cell cycle in intact plant tissues. *Science* **1983**, *220*, 1049–1051. [CrossRef]

76. Penksza, K. *Festuca pseudovaginata*, a new species from sandy areas of the Carpathian Basin. *Acta Bot. Hung.* **2003**, *45*, 365–372. [CrossRef]

77. Lisztes-Szabó, Z.; Kovács, S.; Balogh, P.; Daróczi, L.; Penksza, K.; Pető, Á. Quantifiable differences between phytolith assemblages detected at species level: Analysis of the leaves of nine Poa species (*Poaceae*). *Acta Soc. Bot. Pol.* **2015**, *84*, 369–383. [CrossRef]

78. Horánszky, A. *Festuca*-tanulmányok I. [*Festuca*-studies I.]. *Bot. Közlemények* **1969**, *56*, 149–154.

79. Horánszky, A. *Festuca*-tanulmányok II. [*Festuca*-studies II.]. *Bot. Közlemények* **1970**, *57*, 207–215.

80. Ladányi, Z.; Rakonczai, J.; Kovács, F.; Geiger, J.; Deák, J.Á. The effect of recent climatic change on the Great Hungarian Plain. *Cereal Res. Commun.* **2009**, *37*, 477–480.

81. Ladányi, Z.; Rakonczai, J.; Van Leeuwen, B. Evaluation of precipitation-vegetation interaction on a climate-sensitive landscape using vegetation indices. *J. Appl. Remote Sens.* **2011**, *5*, 53519. [CrossRef]

82. Jones, L.H.P.; Handreck, K.A. Studies of silica in the oat plant. *Plant Soil* **1965**, *23*, 79–96. [CrossRef]

83. Madella, M.; Jones, M.; Echlin, P.; Powers-Jones, A.; Moore, M. Plant water availability and analytical microscopy of phytoliths: Implications for ancient irrigation in arid zones. *Quat. Int.* **2009**, *193*, 32–40. [CrossRef]

84. Raven, J.A. The transport and function of silicon in plants. *Biol. Rev.* **1983**, *58*, 179–207. [CrossRef]

85. Sangster, A.G.; Parry, D.W. Silica Deposition in the Grass Leaf in Relation to Transpiration and the Effect of Dinitrophenol. *Ann. Bot.* **1971**, *35*, 667–677. [CrossRef]

Publisher's Note: MDPI stays neutral with regard to jurisdictional claims in published maps and institutional affiliations.

 *forests*

Article

Wood Pastures: A Transitional Habitat between Forests and Pastures for Dung Beetle Assemblages

László Somay [1,2,3], Viktor Szigeti [1], Gergely Boros [1,3], Réka Ádám [4] and András Báldi [1,2,*]

[1]   Lendület Ecosystem Services Research Group, Centre for Ecological Research, Institute of Ecology and Botany, 2-4 Alkotmány utca, 2163 Vácrátót, Hungary; somay.laszlo@ecolres.hu (L.S.); szigeti.viktor@ecolres.hu (V.S.); boros.gergely@szie.hu (G.B.)

[2]   GINOP Ecosystems Group, Centre for Ecological Reseach, 3 Klebelsberg Kuno utca, 8237 Tihany, Hungary

[3]   Department of Zoology and Ecology, Szent István University, 1 Páter Károly utca, 2100 Gödöllő, Hungary

[4]   Centre for Ecological Reseach, Institute of Ecology and Botany, 2-4 Alkotmány utca, 2163 Vácrátót, Hungary; helleborus42@gmail.com

*   Correspondence: baldi.andras@ecolres.hu

Received: 24 November 2020; Accepted: 22 December 2020; Published: 28 December 2020

Abstract: Wood pastures are home to a variety of species, including the dung beetle. Dung beetles are an important functional group in decomposition. Specifically, in terms of livestock manure, they not only contribute to nutrient cycling but are key players in supporting human and animal health. Dung beetles, however, are declining in population, and urgent recommendations are needed to reverse this trend. Recommendations need to be based on solid evidence and specific habitats. Herein, we aimed to investigate the role of an intermediate habitat type between forests and pastures. Wood pastures are key areas for dung beetle conservation. For this reason, we compared dung beetle assemblages among forests, wood pastures, and grasslands. We complemented this with studies on the effects of dung type and season at three Hungarian locations. Pitfall traps baited with cattle, sheep, or horse dung were used in forests, wood pastures, and pasture habitats in spring, summer, and autumn. Dung beetle assemblages of wood pastures showed transient characteristics between forests and pastures regarding their abundance, species richness, Shannon diversity, assemblage composition, and indicator species. We identified a strong effect of season and a weak of dung type. Assemblage composition proved to be the most sensitive measure of differences among habitats. The conservation of dung beetles, and the decomposition services they provide, need continuous livestock grazing to provide fresh dung, as well as the maintenance of wood pastures where dung beetle assemblages typical of forests and pastures can both survive.

Keywords: abundance; community composition; decomposition; Shannon diversity; ecosystem service; Geotrupidae; grassland; indicator species; land use; Scarabaeidae; species richness

---

1. Introduction

Every year, four billion tons of livestock feces are produced in Europe [1]. Its decomposition is an essential function of ecosystems, and dung beetles (Scarabaeidae: Scarabaeinae, Aphodiinae; Geotrupidae), which are widespread across most terrestrial habitats, play an important role in it [2,3]. Therefore, the conservation of dung beetles and the sustenance of the ecosystem service they provide is an important task. Dung beetles are involved in the decomposition of feces, not only by means of consumption by both larvae and imagoes [4] but even more importantly by chopping, spreading, and burying it, thereby making it available for other decomposing organisms [5,6]. By accelerating decomposition, they contribute to the soil's nutrient cycle and stimulate plant growth by fertilization (increasing the nutrient content of the soil) [7]. Their vertical tunnels promote the mixing of soil layers, increase aeration, and permeability [8].

They also have an important role in maintaining human and animal health by removing the feces of domestic animals from the surface, thus controlling the number of parasitic worms and dung flies [9]. In addition, they assist the secondary distribution of plant seeds present in dung [10]. All of these activities are considered as valuable ecosystem services [3], resulting in significant financial savings for agriculture [11,12].

Only sufficiently diverse assemblages and abundant populations of dung beetles are able to perform their ecological function (acknowledging that body size and functional groups are also important factors [13,14]), but they are declining worldwide. The most serious adverse effects are land use changes, habitat fragmentation [15,16], and declining wild mammal species that provide feces [17,18], all of which transform and intensify traditional animal husbandry [19]. Overgrazing or abandoning pastures leads to a decrease in both abundance and species diversity [20–22]. The negative effects of agricultural intensification, and increased use of pesticides and insecticides, have already been demonstrated on a number of arthropod taxa [23], including dung beetles [24]. Numerous studies have shown that veterinary medical products, widely used in animal health today, and their breakdown products can cause drastic reductions in dung beetle populations as they appear in pastures through feces [14,25,26].

Dung beetles' lifestyles and ecological needs make them sensitive to environmental changes and they are therefore an important indicator group used in more and more studies that assess the state of the environment worldwide [27]. Farming and nature conservation practices that sustain species-rich communities able to perform their ecological functions are essential to the successful conservation of dung beetles. There are, however, still significant knowledge gaps. For example, both grassland and forest dung beetles are well-studied globally, which is important as dung beetle assemblages of forests and grasslands. These habitats are different, e.g., in terms of total biomass, species and functional compositions, and their functions in dung removal [15]. Buse and Entling [28] also found that dung removal is significantly different between forests and grasslands. These are in line with the evidence that dung beetles are highly sensitive to habitat modifications [16] and habitat structure at the local and landscape scales [29,30]. There is, however, a unique European habitat, the wood pasture, a transition between forest and grassland, where trees are scattered across pastures [31]. Wood pastures are still present in the Mediterranean and in Eastern Europe as part of farming systems [32], and present a socio-ecological framework for sustainable agriculture with high biodiversity. Such a unique habitat may provide novel information on how dung beetle assemblages form.

In Europe, including Hungary, the biggest threat to dung beetles is the decline of traditional farming systems. It has changed in many ways since the middle of the 20th century [33,34]. Grazing livestock has been significantly reduced, leaving many former pastures abandoned, while extensive animal husbandry was mostly replaced by intensive livestock farming. Although wood pastures and forest grazing were also well-established as important traditional practices in the last century [35], today only open grasslands are grazed. Nowadays, only a few wood pastures are used actively, but more and more are being restored and grazed mainly for nature conservation purposes. Forest grazing in Hungary has gradually declined and was finally completely ceased in the 1960s [35]. All these trends changed the resources for dung beetles, making them a threatened, declining group.

The objective of our study was to determine the effects of habitat type, dung type, and season on species richness, abundance, diversity, and species composition of dung beetle assemblages. Our aims were as follows:

First, we wanted to find out where (i.e., in what habitat) do dung beetles occur? Our study compared three habitat types with different levels of woodland cover: open pastures, wood pastures, and forests. Wood pastures are acknowledged as valuable habitats for several other taxa [36,37], and we were interested in whether (i) they exhibit an outstanding diversity that harbors both forest and grassland species and thereby have their own specific species pool or if (ii) they represent a transitional situation for dung beetles.

Second, we were interested in what other factors influence dung beetle assemblages. A locality effect, for example, would suggest that the generalization of any results and their applicability at other locations are limited. In addition, to analyse the effect of location, we also studied the effect of dung type and season, both of which are known to influence dung beetles [17,30] and thus may provide additional information to understand the distribution of dung beetles across habitat types. In our study, we compared the dung beetle assemblages in the feces of the three most important grazing livestock in Europe: cattle, horse, and sheep taken in different seasons (spring, summer, and autumn).

Third, we intended to provide research evidence to guide nature conservation management of wood pastures for the preservation of dung beetle assemblages and this unique habitat. Although some such results were known already from other regions or continents, management, which is context dependent, including locality, cannot directly use those results.

2. Materials and Methods

2.1. Study Area

Our study was conducted in three predominantly forested landscapes in the Northern and Transdanubian Mountains of Hungary (sampling locations are given as KMZ (Keyhole Markup language Zipped) files for Google Earth in the Supplementary Materials). The selected pastures and wood pastures were actively grazed. In the forests, there was no livestock grazing; however, wild ungulates such as red deer, roe deer, and wild boar were common. The sites were selected to meet two main criteria. First, they were to include patches of actually grazed wood pastures, as it is the rarest habitat type. Second, there had to be both actively used pastures and seminatural forests in the vicinity (within 2.5 km).

2.1.1. Cserépfalu

The site is a ca. 100 ha area of grasslands and wood pastures, surrounded by forests within the Bükk National Park to the northeast of Cserépfalu at the altitude of 250–350 m above sea level. Two-thirds of the area is wood pasture, while one-third is open pasture. On the open pastures and in the lower parts of the wood pasture, ca. 40 Hungarian grey cattle have been grazed for some years before the time of our sampling. In the year preceding the study, about 200 sheep and some goats were introduced to the wood pasture and herded across the area two or three times a week.

2.1.2. Hollókő

The site is at 250–300 m above sea level within the Hollókő Landscape Protection Area, mostly to the west of the village of Hollókő. In Hollókő, as part of a habitat reconstruction program completed in 2013, nearly 20 hectares of afforested land were cleared and the former wooded pasture was restored. The relatively small wood pasture on the western slope of the Várhegy is completely surrounded by forests. Pastures and hay meadows are situated in larger blocks around the village. On the large western pastures, a herd of ca. 100 head of cattle has been grazed. These animals were herded through the wood pastures for a few times a year for short periods. One paddock on the grassland bordering the southeastern part of the village has been used to keep about 30 sheep and a couple of horses.

2.1.3. Balatonakali-Dörgicse

The site is between the villages of Balatonakali and Dörgicse, 145–240 m above sea level. The ca. 200 ha wood pasture is part of the Natura 2000 network. The western part of the area was cleared a few years ago and sparse wooded vegetation was left behind. The open pasture with a sheep paddock stretches more or less continuously along the southern border of the village of Dörgicse. The southwestern and northwestern parts of the area are covered by continuous forests. In the sheep paddock near Dörgicse, about 1000 head of sheep are kept. They are grazed in smaller flocks mostly on the open pasture and the nearest parts of the wood pasture.

2.2. Sampling Design and Dung Beetle Trapping

The dung beetle assemblages have been sampled three times in 2016, in spring (May), summer (July), and autumn (October). Sampling was carried out simultaneously in all three sites. In total, 81 pitfall trap units baited with either cattle, horse, or sheep dung (three of each for every habitat in all three locations) were installed. The pitfall traps baited with different dungs were placed as if they were vertices of a triangle with sides slightly more than 10 m. The location of the sampling plots was selected to meet two basic criteria: they should be at least 100 m away from each other [38] and the edge of the habitat type should be at least 50 m away. As far as the terrain allowed, traps were installed in locations sheltered by stumps, smaller trees, or shrubs to avoid damage caused by trampling.

The 1 L containers (diameter: 11 cm, height: 15 cm) were dug into the soil up to their rim. About 3 dl of diluted (50%) propylene-glycol was used as a preservative in each container. The mouth was covered by a hexagonal chicken wire (with a mesh diameter of 25 mm) that was fixed to the soil by U-shaped wires. The dung bait was wrapped in mosquito net and fixed to the wire mesh covering the mouth by bailing wire.

The dung used as bait was always collected freshly for each sampling period from animals grazing on the specific sites. The same batch of dung was used in each trap of a particular site. Livestock were not treated with anthelmintics prior to dung collection. The collected dung was portioned into 400 (cattle, horse) and 100 g (sheep) packs within laboratory conditions. Packs were sterilised by freezing at $-20\,°C$ for at least 72 h. The pitfall traps were emptied after one week. In one case, the container was dug up by animals; this sample was excluded from the analysis. Dung beetles were identified at the species level in the laboratory. Nomenclature follows the Catalogue of Palaearctic Coleoptera [39].

2.3. Data Analysis

We separately tested the effect of four explanatory variables (habitat, season, dung, and locality) on abundance and species richness by generalised linear mixed models (GLMMs; [40]) and on Shannon diversity using linear mixed models (LMMs), where the response variables were the number of species (with Poisson distribution), the abundance (with negative binomial distribution), or the calculated Shannon diversity (with Gaussian distribution) of dung beetles. Replicates (sampling plots with three pitfall traps) were included as random factors. We compared the models fitted with explanatory and random factors to null models (including the random factor only) by ANOVA [41].

Our main aim was to reveal the differences in dung beetle assemblages of forest, wood pasture, and pasture habitat types. Thus, first, we compared species richness, abundances, and Shannon diversity among habitat types by a pairwise comparison with Tukey corrections, based on models described above. Second, we created subsets of season and dung type (thereby creating 9 datasets both for abundances and for number of species) and separately analysed the effect of habitat type. We applied GLMMs and LMMs [40], where the response variables were the number of species, the abundance, or the diversity of dung beetles, while the explanatory variable was only the habitat type (forest, pasture, and wood pasture). Replicates within the locality were included as random factors, and we repeated these analyses for all 9 subsets (3 seasons × 3 dung types) of our dataset. We applied pairwise comparisons with Tukey corrections between habitat types based on model results.

We explored the differences between the composition of dung beetle assemblages with nonmetric multidimensional scaling (NMDS) using a Bray–Curtis similarity measure [42]. We applied a Permutational Multivariate Analysis of Variance (PERMANOVA) [42] to analyse the effect of habitat, season, locality, and dung type as explanatory variables on assemblage composition. We evaluated the association of dung beetle species to habitat, dung type, season, and locality separately, by indicator species analysis [42]. The indicator values of the species were tested via the Monte-Carlo simulation using 10,000 permutations. The accepted significance level was $p < 0.05$.

The statistical analyses were carried out using R 3.4.4 statistical environment [43]. We used "lme4" 1.1-17 and "MASS" 7.3-51.4 packages for models [44,45], "lsmeans" 2.27-62 and "multcomp" 1.4-10 packages for pairwise comparison [46,47], "vegan" 2.5-2. package for nonmetric multidimensional scaling [48], and "labdsv" ver. 1.8-0 for indicator species analysis [49].

3. Results

Altogether, we recorded 57 species and 78,131 individuals across the three locations (Cserépfalu: 48 species, 32,490 individuals, Hollókő: 46 species, 23,611 individuals, and Balatonakali: 44 species, 22,030 individuals). Three species of the family Geotrupidae, 33 species of the subfamily Aphodiinae, and 21 species of the subfamily Scarabaeinae were recorded (Appendix A Table A1).

The most numerous was the Scarabaeinae subfamily (43,209 individuals, 55.3%), then the Aphodiinae subfamily (27,314 individuals, 35.0%), and finally the Geotrupidae family (7608 individuals, 9.7%). The most dominant species was *Nimbus obliteratus*, with 16,540 individuals collected, which was 21% of all individuals. In total, 18 species accounted for nearly 95% of the total abundance, while 39 species accounted for less than 1% (Appendix A Table A1).

Our models did not reveal a significant effect of locality on species richness ($p = 0.116$), abundance ($p = 0.383$), or diversity ($p = 0.145$), while habitat had an effect on diversity ($p = 0.001$), but no effect on species number ($p = 0.070$) nor abundance ($p = 0.302$). Both dung type ($p(\text{sp}) = 0.024$, $p(\text{abu}) = 0.001$) and season ($p(\text{sp}) < 0.001$, $p(\text{abu}) = 0.022$, $p(\text{div}) < 0.001$) had a significant effect. However, we did not find an effect of dung type on diversity ($p = 0.473$).

Considering habitat types, 32,917 individuals of 41 species were recorded from forests, 20,724 individuals of 47 species from wood pastures, and 24,490 individuals of 53 species from grasslands. Species richness and Shannon diversity were lower in forests than in pastures. In the case of wood pastures, these values were between the two distinct habitats. There was no difference in abundances between the three habitats (Figure 1).

Pairwise comparisons of species richness, abundances, and Shannon diversities between habitat types revealed several significant differences between forests and pastures, while wood pastures exhibited an intermediate position (e.g., species richness for sheep dung during spring and summer), except for one case (spring abundance on cattle dung) where wood pasture dung beetle number was lower than in both habitats (Figure 2, Appendix A Table A2).

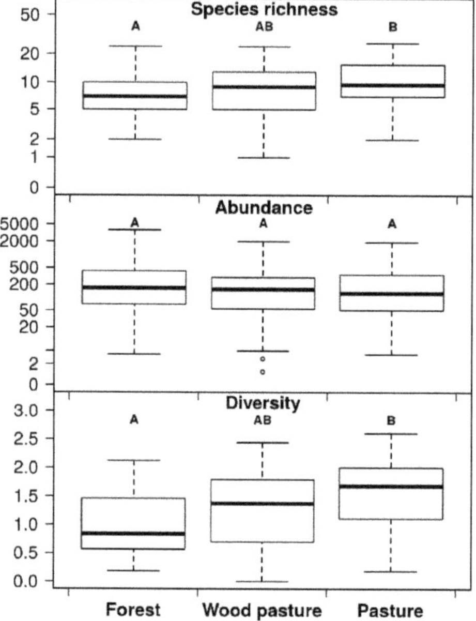

Figure 1. Summarised dung beetle species richness, abundance, and Shannon diversity per habitat types. Box plots show medians (thick line), lower, and upper quartiles (boxes), whiskers include the range of distribution without outliers. Different letters above the boxes show the significant differences between habitats.

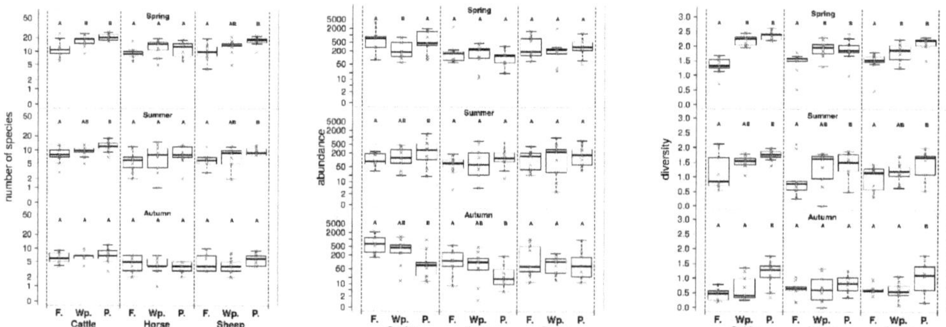

Figure 2. Habitat effect on dung beetle species richness, abundance, and diversity separately for dung type (columns) and seasons (rows). Box plots show medians (thick line), lower, and upper quartiles (boxes), whiskers include the range of distribution without outliers. Different letters above the boxes show the significant differences between habitats within a given season and dung type, applying generalised linear mixed models (GLMMs), linear mixed models (LMMs), and their pairwise comparisons. F.: forest, Wp.: wood pasture, and P.: pasture.

For composition of dung beetle assemblages, the seasonal distinction was the most noticeable effect according to the NMDS analysis (Figure 3A, PERMANOVA test: $R^2 = 0.237$, $p = 0.001$). There was no overlap between forest and grassland dung beetle samples, but wood pastures had an intermediate position between the two other habitat types (Figure 3B, PERMANOVA test: $R^2 = 0.109$, $p = 0.001$). Furthermore, assemblage composition showed significant differences among the three localities in Hungary (Figure 3C, PERMANOVA test: $R^2 = 0.081$, $p = 0.001$), but the effect of locality was smaller compared to the effect of season and habitat. Dung type had no significant effect on the composition of dung beetle assemblages (Figure 3D, PERMANOVA test: $R^2 = 0.017$, $p = 0.322$).

The analysis of indicator species regarding habitat, dung type, and season showed that 12 species were linked to pastures, 6 to forests, and only one to wood pastures (Table 1). Only cattle feces had three indicator species, while the other two dung types had none. Spring had 20 indicator species, while summer and autumn had less than half of that (Table 1).

Table 1. Dung beetle species with significant indicator values for habitat, dung type, and season.

Species	Indicator Value	*p*-Value	Habitat
Anoplotrupes stercorosus	0.664	0.000	Forest
Melinopterus pubescens	0.251	0.011	Forest
Onthophagus coenobita	0.606	0.002	Forest
Plagiogonus arenarius	0.341	0.028	Forest
Planolinus fasciatus	0.404	0.000	Forest
Volinus sticticus	0.778	0.000	Forest
Caccobius schreberi	0.398	0.002	Pasture
Chilothorax distinctus	0.249	0.003	Pasture
Colobopterus erraticus	0.290	0.001	Pasture
Euoniticellus fulvus	0.642	0.000	Pasture
Melinopterus consputus	0.294	0.022	Pasture
Onthophagus illyricus	0.628	0.000	Pasture
Onthophagus medius	0.252	0.021	Pasture
Onthophagus ovatus	0.655	0.000	Pasture
Onthophagus ruficapillus	0.606	0.000	Pasture
Onthophagus taurus	0.665	0.000	Pasture
Onthophagus vacca	0.259	0.001	Pasture
Otophorus haemorrhoidalis	0.311	0.002	Pasture
Onthophagus joannae	0.248	0.003	Wood pasture
Species	**Indicator Value**	***p*-Value**	**Dung Type**
Aphodius fimetarius	0.287	0.002	Cattle
Otophorus haemorrhoidalis	0.238	0.025	Cattle
Teuchestes fossor	0.185	0.009	Cattle
Species	**Indicator Value**	***p*-Value**	**Season**
Acrossus depressus	0.333	0.000	Spring
Acrossus luridus	0.630	0.000	Spring
Colobopterus erraticus	0.264	0.001	Spring

Table 1. *Cont.*

Species	Indicator Value	*p*-Value	Habitat
Copris lunaris	0.763	0.000	Spring
Esymus pusillus	0.922	0.000	Spring
Eudolus quadriguttatus	0.444	0.000	Spring
Euoniticellus fulvus	0.411	0.002	Spring
Euorodalus paracoenosus	0.593	0.000	Spring
Onthophagus coenobita	0.941	0.000	Spring
Onthophagus illyricus	0.607	0.001	Spring
Onthophagus joannae	0.204	0.014	Spring
Onthophagus lemur	0.573	0.000	Spring
Onthophagus medius	0.481	0.000	Spring
Onthophagus vacca	0.163	0.043	Spring
Onthophagus verticicornis	0.960	0.000	Spring
Oxyomus sylvestris	0.454	0.000	Spring
Phalacronothus biguttatus	0.208	0.006	Spring
Plagiogonus arenarius	0.782	0.000	Spring
Sisyphus schaefferi	0.672	0.000	Spring
Volinus sticticus	0.604	0.000	Spring
Acanthobodilus immundus	0.200	0.033	Summer
Bodilus lugens	0.259	0.001	Summer
Caccobius schreberi	0.327	0.017	Summer
Geotrupes spiniger	0.633	0.000	Summer
Onthophagus grossepunctatus	0.471	0.012	Summer
Onthophagus taurus	0.444	0.036	Summer
Trypocopris vernalis	0.701	0.000	Summer
Chilothorax distinctus	0.249	0.004	Autumn
Chilothorax paykulli	0.222	0.002	Autumn
Melinopterus consputus	0.593	0.000	Autumn
Melinopterus prodromus	0.911	0.000	Autumn
Nimbus obliteratus	0.926	0.000	Autumn
Nobius serotinus	0.222	0.004	Autumn
Planolinus fasciatus	0.443	0.000	Autumn
Sigorus porcus	0.148	0.029	Autumn

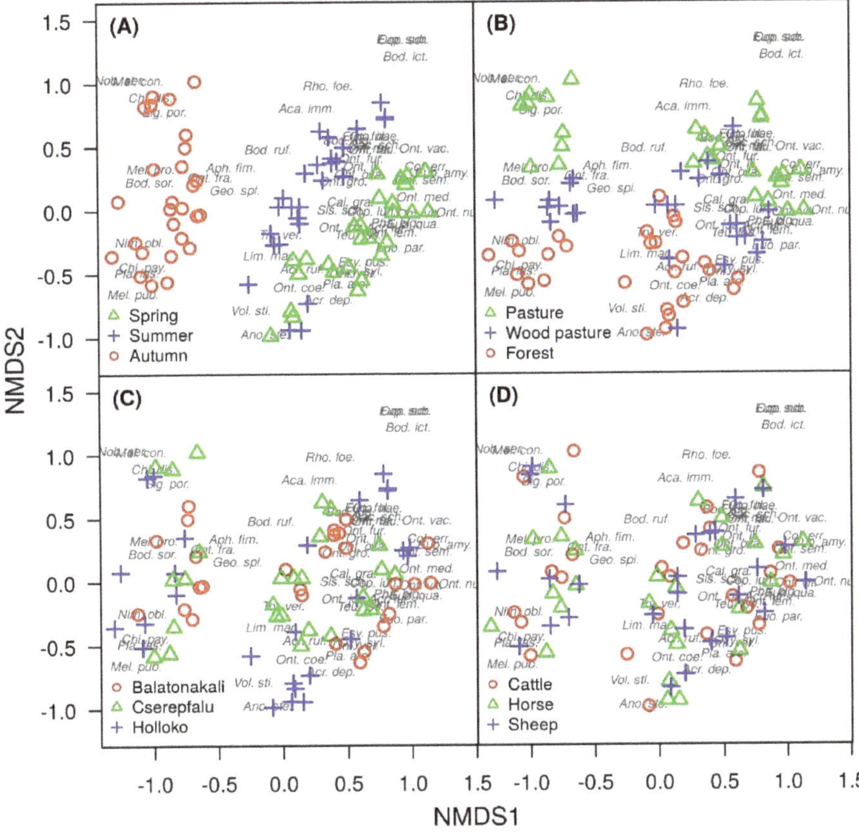

Figure 3. Assembly composition of dung beetles according to (**A**) season, (**B**) habitat, (**C**) locality, and (**D**) dung type, applying nonmetric multidimensional scaling (NMDS).

4. Discussion

Our main question concerned whether wood pastures have distinct dung beetle assemblages compared to forests and pastures. Using various types of analyses to process data on around 80,000 individuals, we found that assemblages in wood pastures showed an intermediate or transient position between the assemblages of forests and pastures, which were distinct. In Central Europe, the majority of dung beetle species inhabit open or semiopen grasslands, and there are fewer woodland species [50] (Scarabaeinae: mostly grassland species, Geotrupidae: mostly forest species, Aphodiinae: both forest and grassland species [15]). Our study showed similar patterns. Dung beetle assemblages on wood pastures showed an intermediate position between forest and pasture assemblages in composition, species richness, and Shannon diversity. In general, this position was present in all but one comparison when broken down according to seasons and dung types. This transient position is also supported by the distribution of the indicator species, wherein we found only one dung beetle (*Onthophagus joannae*) species specific for the wood pastures from the total of 57 dung beetle species and total of 19 habitat indicator species. This pattern in assemblage characteristics probably had an effect on the decomposition function of dung beetles, as Buse and Entling [28] showed.

The role of habitat in structuring dung beetle assemblages is well-known. Several studies found differences in various parameters between forest and pasture assemblages. Ríaz-Díaz et al. [51] found significant differences in species composition between forest and grassland habitats. Jugovic et al. [52] found significant differences in species composition, species richness, and abundance among grazed and overgrown sites. Numa et al. [53] found significant differences for species composition, species dominance, and abundance between sites with wild versus livestock herbivory. All of these studies showed a strong effect of habitat on the structure of dung beetle assemblages. In our case, we also found significant differences between forests and pastures in species richness and Shannon diversity, but not for wood pastures and forests or wood pastures and pastures. This pattern is remarkable, as based on the literature. We may expect three different assemblages in the three different habitats. Similar studies, where forest, pasture, and a transitional habitat (usually shrubland) were involved, found significant differences between habitats, indicating that neither was in a transitional position [29,54]. It seems that this is the general pattern, as this was found for other taxa as well. For example, spider [37] and ground-dwelling beetle assemblages [55] of open pasture habitats and scattered trees (and other habitats) were statistically distinct. Lövei et al. [56] categorised carabids to the forest edge and matrix species as these habitats had distinct assemblages. The transitional position of wood pastures between forests and pastures is interesting, if compared to forest, forest edge, and grassland patterns. Such studies showed that all three habitats had different assemblages for carabids [57], ants and spiders [58], and millipedes [59]. Therefore, the habitat effect in the forest, wood pasture, and pasture system seems to be rather different from forest, edge, and pasture (or grassland) systems, highlighting the uniqueness of the former system.

Our second question was on the potential effects of locality, dung type, and season on assemblages. We found that dung beetle assemblages are similar across three distant locations in Hungary for all studied measures, namely for species number, abundance, Shannon diversity, and composition. The similarity of the three localities is probably linked to the overwhelming role of locally relevant factors in shaping dung beetle assemblages (and decomposition in general); thus, differences between habitat or management types are expected to be larger than differences between locations [60,61]. This similarity of dung beetle assemblages across locations indicates that our results are relevant to other locations (at least in the Pannonian region), which supports the wider applicability of our findings. The habitat effect and lack of locality effect on dung beetle assemblages are probably linked to vegetation characteristics. Several studies showed that vegetation structure (e.g., plant diversity [62]), fine-scale heterogeneity in grazing intensity [63], or vegetation cover [64] highly influenced dung beetle assemblages. Our results on habitat and locality can be explained on the assumption that differences between forests or pastures or wood pastures are smaller across locations than among the three habitat types within a given location.

The majority of Palearctic dung beetle species are considered to be generalists, feeding on various types of dung available. This suggests that a dung type effect on assemblage structure is likely weak. Indeed, we could not clearly identify distinct dung beetle assemblages across the three dung types, which is supported also by a low number of indicator species associated with each dung type. However, we found a significant effect of dung type on abundance and species number, indicating that even generalist species can show different preferences if they have a choice of several dung types [65].

Season seems to be a profound factor in shaping dung beetle assemblages. We already know from several studies that dung beetle species appear in a determined phenological order during the year [52,66,67]. We found a clear distinction in assemblage composition in spring, summer, and autumn for all assemblage measures. Indicator species are also in line with former results [68]. In autumn, the wood pasture samples were more distinct from forests and pastures, while both the spring and the summer samples showed a greater similarity among dung beetle assemblages in the three habitats, suggesting that the seasonality influenced the effect of habitats on dung beetles.

Dung beetle assemblage composition was different in this study for habitats, seasons, and dung types. It seems that composition is a sensitive measure of differences in dung beetle assemblages across a range of conditions [51,69].

The third aim of this study was to provide information for nature conservation management. Dung beetle assemblages of Hungary and the Pannonian biogeographic region are still understudied. During this one-year sampling in three locations, nearly half of the species of Hungarian dung beetle fauna were recorded; this seems to be outstanding for the results of the mosaic-like structural complexity of the studied habitats. The conservation of dung beetles should focus on maintaining compositional diversity, as different species have different roles, and the loss of a species cannot be compensated by the presence of others [70]. Species-specific thermal tolerance is very important [71], as is poor flight capacity limiting dispersal that potentially leads to local extinctions if habitats are further degraded (e.g., Storck-Tonon et al. [72]). The provisioning of fresh dung throughout the year—irrespective of what livestock species produces it—is of major importance for all habitats. The manure of wild ungulates is not sufficient to ensure the conservation of the regional dung beetle species pool [73]. From a practical point of view, this means that if we want to maintain species-rich communities and the associated ecosystem services, we can do so by increasing the availability of the right amount and quality of manure throughout the year. Extensive animal husbandry (i.e., grazing livestock from spring to late autumn) is one way to enhance the habitat for dung beetles. It is imperative for the conservation of diverse assemblages that grazing take place in the long term. Grazing continuity has a positive effect on species numbers both for generalists and specialists [74]. The use of veterinary medical products needs to be reconsidered to avoid harm to dung beetles [14]. Long-term conservation needs to consider climate change effects and connectivity of suitable dung beetle habitats to provide dispersal possibilities. Active conservation planning is needed, as only sufficiently stable and abundant populations as well as diverse assemblages of dung beetles are able to perform their ecological functions that benefit people [15,75].

5. Conclusions

We concluded that wood pastures are key habitats for dung beetle conservation, as they harbor dung beetle assemblages that show a transition between forest and grassland assemblages. Wood pastures should not be viewed as a refuge for all forest species [76], but as a transitional habitat where some forest and pasture species can be conserved into one habitat. Therefore, wood pastures must be maintained and their characteristically large trees need to be conserved and replaced if needed. Proper management by grazing livestock is also essential not only for dung beetles but for various ecologically important taxa. The conservation of such ecosystems should be a priority. Throughout Europe, there are already several completed or ongoing projects aimed at the restoration and conservation management of wood pastures. Such actions should be funded more widely for the benefits they provide both in terms of their unique biodiversity and local livelihoods not only in Europe, but worldwide.

Supplementary Materials: The following are available online at http://www.mdpi.com/1999-4907/12/1/25/s1, Sampling locations (KMZ file).

Author Contributions: Conceptualization, L.S.; methodology, L.S.; field and lab work, L.S. and R.Á.; data curation, L.S., R.Á., and V.S.; formal analysis, V.S.; writing—original draft preparation, L.S., V.S., G.B., R.Á. and A.B.; writing—review and editing, L.S., V.S., G.B., R.Á. and A.B.; supervision, A.B.; funding acquisition, A.B. All authors have read and agreed to the published version of the manuscript.

Funding: This research was funded by the Doctoral School of Environmental Sciences of the Szent István University, the "Sustainable use of ecosystem services—research for mitigating the negative effect of climate change, land use change and biological invasion" project (GINOP-2.3.2-15-2016-00019), and partly by the EU H2020 project SUPER-G (https://www.super-g.eu, grant agreement N. 774124). V.S. was funded by the National Research, Development and Innovation Office (FK 123813).

Acknowledgments: We would like to thank the two anonymous reviewers for their valuable comments. We would also like to thank Krisztián Harmos (Bükk National Park) for field assistance, as well as Brigitta Palotás and Anikó Zölei for their help in preparing the paper.

Conflicts of Interest: The authors declare no conflict of interest.

Appendix A

Table A1. List of dung beetle species, and their abundances according to habitats at the three regions in Hungary (B—Balatonakali, H—Hollókő, C—Cserépfalu).

	Abbr.	Forest				Wood Pasture				Pasture				Total	%
		B	H	C	Σ	B	H	C	Σ	B	H	C	Σ		
Geotrupidae															
Anoplotrupes stercorosus (Scriba, 1791)	Ano_ste	1	1671	117	**1789**	0	13	90	**103**	0	3	2	**5**	1897	2.43
Geotrupes spiniger (Marsham, 1802)	Geo_spi	29	9	9	**47**	18	1	13	**32**	23	6	16	**45**	124	0.16
Trypocopris vernalis (Linnaeus, 1758)	Try_ver	1534	156	914	**2604**	1741	148	709	**2598**	345	21	19	**385**	5587	7.15
Scarabaeinae															
Caccobius schreberi (Linnaeus, 1767)	Cac_sch	0	0	0	**0**	442	10	0	**452**	685	231	7	**923**	1375	1.76
Copris lunaris (Linnaeus, 1758)	Cop_lun	11	1	15	**27**	14	1	49	**64**	33	12	74	**119**	210	0.27
Euoniticellus fulvus (Goeze, 1777)	Euo_ful	0	0	0	**0**	26	6	46	**78**	73	163	274	**510**	588	0.75
Euonthophagus amyntas (Olivier, 1789)	Euo_amy	0	0	0	**0**	4	0	0	**4**	11	0	0	**11**	15	0.02
Onthophagus coenobita (Herbst, 1783)	Ont_coe	321	294	1159	**1774**	31	326	136	**493**	11	78	29	**118**	2385	3.05
Onthophagus fracticornis (Preyssler, 1790)	Ont_fra	15	20	1451	**1486**	176	224	470	**870**	288	103	836	**1227**	3583	4.59
Onthophagus furcatus (Fabricius, 1781)	Ont_fur	0	0	0	**0**	0	0	3	**3**	0	0	4	**4**	7	0.01
Onthophagus grossepunctatus Reitter, 1905	Ont_gro	226	0	157	**383**	2301	6	799	**3106**	1504	281	564	**2349**	5838	7.47
Onthophagus illyricus (Scopoli, 1763)	Ont_ill	11	2	41	**54**	153	70	164	**387**	372	1309	767	**2448**	2889	3.70
Onthophagus joannae Goljan, 1953	Ont_joa	0	0	0	**0**	0	10	41	**51**	0	0	10	**10**	61	0.08
Onthophagus lemur (Fabricius, 1781)	Ont_lem	20	0	10	**30**	207	0	2	**209**	111	1	3	**115**	354	0.45
Onthophagus medius (Kugelann, 1792)	Ont_med	0	2	0	**2**	33	12	0	**45**	460	1190	1	**1651**	1698	2.17
Onthophagus nuchicornis (Linnaeus, 1758)	Ont_nuc	0	0	0	**0**	0	0	0	**0**	1	0	0	**1**	1	0.00
Onthophagus ovatus (Linnaeus, 1767)	Ont_ova	2	10	3	**15**	63	189	357	**609**	96	2614	617	**3327**	3951	5.06
Onthophagus ruficapillus Brulle, 1832	Ont_ruf	1	0	2	**3**	403	6	15	**424**	1436	1049	160	**2645**	3072	3.93
Onthophagus semicornis (Panzer, 1798)	Ont_sem	0	0	0	**0**	0	0	0	**0**	0	1	1	**2**	2	0.00
Onthophagus taurus (Schreber, 1759)	Ont_tau	1	1	2	**4**	49	17	9	**75**	179	1078	98	**1355**	1434	1.84
Onthophagus vacca (Linnaeus, 1767)	Ont_vac	0	0	0	**0**	0	0	0	**0**	9	7	1	**17**	17	0.02
Onthophagus verticicornis (Laicharting, 1781)	Ont_ver	1003	101	5945	**7049**	89	81	1111	**1281**	95	448	475	**1018**	9348	11.96
Onthophagus vitulus (Fabricius, 1777)	Ont_vit	0	0	0	**0**	1	0	0	**1**	0	0	0	**0**	1	0.00
Sisyphus schaefferi (Linnaeus, 1758)	Sis_sch	751	8	796	**1555**	713	49	1943	**2705**	943	642	535	**2120**	6380	8.17
Aphodiinae															
Acanthobodilus immundus (Creutzer, 1799)	Aca_imm	0	1	0	**1**	1	1	0	**2**	0	27	0	**27**	30	0.04
Acrossus depressus (Kugelann, 1792)	Acr_dep	1	4	20	**25**	0	1	1	**2**	0	0	3	**3**	30	0.04
Acrossus luridus (Fabricius, 1775)	Acr_lur	0	1	8	**9**	7	88	23	**118**	7	276	39	**322**	449	0.57

Table A1. Cont.

		Forest				Wood Pasture				Pasture				Total	%
	Abbr.	B	H	C	Σ	B	H	C	Σ	B	H	C	Σ		
Acrossus rufipes (Linnaeus, 1758)	Acr_ruf	0	2	27	29	0	0	14	14	0	0	7	7	50	0.06
Aphodius fimetarius (Linnaeus, 1758)	Aph_fim	0	0	4	4	0	2	5	7	1	1	20	22	33	0.04
Bodiloides ictericus (Laicharting, 1781)	Bod_ict	0	0	0	0	0	0	0	0	0	7	0	7	7	0.01
Bodilopsis rufa (Moll, 1782)	Bod_ruf	1	0	0	1	0	0	0	0	0	0	1	1	2	0.00
Bodilopsis sordida (Fabricius, 1775)	Bod_sor	0	1	0	1	0	1	0	1	0	3	0	3	5	0.01
Bodilus lugens (Creutzer, 1799)	Bod_lug	0	0	0	0	112	0	0	112	9	2	0	11	123	0.16
Calamosternus granarius (Linnaeus, 1767)	Cal_gra	0	0	1	1	1	0	0	1	0	0	1	1	3	0.00
Chilothorax distinctus (O. F. Muller, 1776)	Chi_dis	0	0	1	1	0	0	0	0	19	3	2	24	25	0.03
Chilothorax paykulli (Bedel, 1908)	Chi_pay	0	0	44	44	1	0	0	1	3	0	0	3	48	0.06
Colobopterus erraticus (Linnaeus, 1758)	Col_err	0	0	0	0	0	2	0	2	6	81	2	89	91	0.12
Coprimorphus scrutator (Herbst, 1789)	Cop_scr	0	0	0	0	0	0	0	0	0	1	0	1	1	0.00
Esymus pusillus (Herbst, 1789)	Esy_pus	3	29	22	54	6	36	69	111	11	174	99	284	449	0.57
Eudolus quadriguttatus (Herbst, 1783)	Euo_qua	0	0	3	3	4	0	5	9	22	0	10	32	44	0.06
Euorodalus paracoenosus (Balthasar and Hrubant, 1960)	Euo_par	2	0	20	22	28	0	49	77	29	13	17	59	158	0.20
Eupleurus subterraneus (Linnaeus, 1758)	Eup_sub	0	0	0	0	0	0	0	0	0	1	0	1	1	0.00
Limarus maculatus (Sturm, 1800)	Lim_mac	0	0	2	2	0	0	0	0	0	0	0	0	2	0.00
Melinopterus consputus (Creutzer, 1799)	Mel_con	0	4	2	6	9	2	1	12	54	1715	446	2215	2233	2.86
Melinopterus prodromus (Brahm, 1790)	Mel_pro	31	22	3	56	60	90	13	163	301	139	23	463	682	0.87
Melinopterus pubescens (Sturm, 1800)	Mel_pub	0	280	1	281	0	9	0	9	0	0	0	0	290	0.37
Nimbus obliteratus (Panzer, 1823)	Nim_obl	1869	2133	6801	10,803	1082	3681	849	5612	66	54	5	125	16,540	21.17
Nobius serotinus (Panzer, 1799)	Nob_ser	0	3	0	3	0	1	0	1	0	155	10	165	169	0.22
Otophorus haemorrhoidalis (Linnaeus, 1758)	Oto_hae	0	0	1	1	0	0	4	4	1	59	9	69	74	0.09
Oxyomus sylvestris (Scopoli, 1763)	Oxy_syl	6	1	7	14	9	0	4	13	1	3	1	5	32	0.04
Phalacronothus biguttatus (Germar, 1824)	Pha_big	1	0	0	1	1	0	3	4	10	1	0	11	16	0.02
Plagiogonus arenarius (Olivier, 1789)	Pla_are	272	70	597	939	275	3	300	578	37	2	74	113	1630	2.09
Planolinus fasciatus (Olivier, 1789)	Pla_fas	554	105	198	857	1	0	7	8	0	0	0	0	865	1.11
Rhodaphodius foetens (Fabricius, 1787)	Rho_foe	0	0	0	0	0	0	0	0	0	0	1	1	1	0.00
Sigorus porcus (Fabricius, 1792)	Sig_por	0	0	0	0	2	0	1	3	0	0	2	2	5	0.01
Teuchestes fossor (Linnaeus, 1758)	Teu_fos	0	1	1	2	0	1	0	1	0	8	0	8	11	0.01
Volinus sticticus (Panzer, 1798)	Vol_sti	49	1593	1293	2935	0	27	242	269	0	10	1	11	3215	4.11
Abundance		6715	6525	19,677	32,917	8063	5114	7547	20,724	7252	11,972	5266	24,490	78,131	100.00
Species number		25	28	35	41	34	32	34	47	35	41	42	53	57	

Table A2. Significance values for the comparisons in Figure 2, on dung beetle abundance and species number according to dung type, season and habitat (* $p \leq 0.05$,** $p \leq 0.01$,*** $p \leq 0.001$).

		Contrast		Adjusted *p*-Value		
				Abundance	Number of Species	Diversity
Cattle	Spring	Forest	Wood pasture	0.0037 **	0.0124 *	<0.0001 ***
		Forest	Pasture	0.9996	0.0001 ***	<0.0001 ***
		Wood pasture	Pasture	0.0053 **	0.3134	0.2063
	Summer	Forest	Wood pasture	0.7217	0.6062	0.1177
		Forest	Pasture	0.0083 **	0.0123 *	0.0055 **
		Wood pasture	Pasture	0.0701	0.1337	0.3470
	Autumn	Forest	Wood pasture	0.4623	0.8890	0.4605
		Forest	Pasture	0.0067 **	0.5829	0.0024 **
		Wood pasture	Pasture	0.1523	0.8565	0.0380 *
Horse	Spring	Forest	Wood pasture	0.2544	0.0764	0.0106 *
		Forest	Pasture	0.0995	0.1767	0.0214 *
		Wood pasture	Pasture	0.8438	0.9165	0.9486
	Summer	Forest	Wood pasture	0.8803	0.4303	0.1982
		Forest	Pasture	0.2188	0.1953	0.0443 *
		Wood pasture	Pasture	0.5154	0.8739	0.7153
	Autumn	Forest	Wood pasture	0.8827	0.9741	0.9830
		Forest	Pasture	0.0344 *	0.6473	0.5260
		Wood pasture	Pasture	0.1058	0.7753	0.6300
Sheep	Spring	Forest	Wood pasture	0.8016	0.1234	0.0473 *
		Forest	Pasture	0.9111	0.0004 ***	0.0001 ***
		Wood pasture	Pasture	0.9834	0.1358	0.0537
	Summer	Forest	Wood pasture	0.4881	0.4314	0.3512
		Forest	Pasture	0.4728	0.0420 *	0.0166 *
		Wood pasture	Pasture	0.9993	0.4548	0.2640
	Autumn	Forest	Wood pasture	0.0843	0.5931	0.9162
		Forest	Pasture	0.4491	0.7658	0.0278 *
		Wood pasture	Pasture	0.6574	0.2209	0.0114 *

References

1.   Berendes, D.M.; Yang, P.J.; Lai, A.; Hu, D.; Brown, J. Estimation of global recoverable human and animal faecal biomass. *Nat. Sustain.* **2018**, *1*, 679–685. [CrossRef]
2.   Hanski, I.; Cambefort, Y. *Dung Beetle Ecology*; Princeton University Press: Princeton, NJ, USA, 1991; ISBN 978-0-691-08739-9.
3.   Nichols, E.; Spector, S.; Louzada, J.; Larsen, T.; Amezquita, S.; Favila, M. Ecological functions and ecosystem services provided by Scarabaeinae dung beetles. *Biol. Conserv.* **2008**, *141*, 1461–1474. [CrossRef]
4.   Holter, P. Herbivore dung as food for dung beetles: Elementary coprology for entomologists. *Ecol. Entomol.* **2016**, *41*, 367–377. [CrossRef]
5.   Tixier, T.; Bloor, J.M.G.; Lumaret, J.-P. Species-specific effects of dung beetle abundance on dung removal and leaf litter decomposition. *Acta Oecol.* **2015**, *69*, 31–34. [CrossRef]
6.   Slade, E.M.; Roslin, T.; Santalahti, M.; Bell, T. Disentangling the 'brown world' faecal-detritus interaction web: Dung beetle effects on soil microbial properties. *Oikos* **2016**, *125*, 629–635. [CrossRef]
7.   Bang, H.S.; Lee, J.-H.; Kwon, O.S.; Na, Y.E.; Jang, Y.S.; Kim, W.H. Effects of paracoprid dung beetles (Coleoptera: Scarabaeidae) on the growth of pasture herbage and on the underlying soil. *Appl. Soil Ecol.* **2005**, *29*, 165–171. [CrossRef]
8.   Brown, J.; Scholtz, C.H.; Janeau, J.-L.; Grellier, S.; Podwojewski, P. Dung beetles (Coleoptera: Scarabaeidae) can improve soil hydrological properties. *Appl. Soil Ecol.* **2010**, *46*, 9–16. [CrossRef]
9.   Ridsdill-Smith, T.J.; Edwards, P.B. Biological Control: Ecosystem Functions Provided by Dung Beetles. In *Ecology and Evolution of Dung Beetles*; Simmons, L.E., Ridsdill-Smith, T.J., Eds.; Wiley-Blackwell: Chichester, UK, 2011; pp. 245–266. ISBN 978-1-4443-4200-0.
10. D'Hondt, B.; Bossuyt, B.; Hoffmann, M.; Bonte, D. Dung beetles as secondary seed dispersers in a temperate grassland. *Basic Appl. Ecol.* **2008**, *9*, 542–549. [CrossRef]
11. Losey, J.E.; Vaughan, M. The Economic Value of Ecological Services Provided by Insects. *BioScience* **2006**, *56*, 311–323. [CrossRef]
12. Beynon, S.A.; Wainwright, W.; Christie, M. The application of an ecosystem services framework to estimate the economic value of dung beetles to the U.K. cattle industry. *Ecol. Entomol.* **2015**, *40*, 124–135. [CrossRef]
13. Tonelli, M.; Verdú, J.R.; Morelli, F.; Zunino, M. Dung beetles: Functional identity, not functional diversity, accounts for ecological process disruption caused by the use of veterinary medical products. *J. Insect Conserv.* **2020**, *24*, 643–654. [CrossRef]
14. Verdu, J.R.; Lobo, J.M.; Sánchez-Piñero, F.; Gallego, B.; Numa, C.; Lumaret, J.-P.; Cortez, V.; Ortiz, A.J.; Tonelli, M.; García-Teba, J.P.; et al. Ivermectin residues disrupt dung beetle diversity, soil properties and ecosystem functioning: An interdisciplinary field study. *Sci. Total Environ.* **2018**, *618*, 219–228. [CrossRef]
15. Frank, K.; Hülsmann, M.; Assmann, T.; Schmitt, T.; Blüthgen, N. Land use affects dung beetle communities and their ecosystem service in forests and grasslands. *Agric. Ecosyst. Environ.* **2017**, *243*, 114–122. [CrossRef]
16. Nichols, E.; Larsen, T.; Spector, S.; Davis, A.L.; Escobar, F.; Favila, M.; Vulinec, K. Global dung beetle response to tropical forest modification and fragmentation: A quantitative literature review and meta-analysis. *Biol. Conserv.* **2007**, *137*, 1–19. [CrossRef]
17. Nichols, E.; Gardner, T.A.; Peres, C.A.; Spector, S. The Scarabaeinae Research Network Co-declining mammals and dung beetles: An impending ecological cascade. *Oikos* **2009**, *118*, 481–487. [CrossRef]
18. Bogoni, J.A.; Da Silva, P.G.; Peres, C.A. Co-declining mammal–dung beetle faunas throughout the Atlantic Forest biome of South America. *Ecography* **2019**, *42*, 1803–1818. [CrossRef]
19. Hutton, S.A.; Giller, P.S. The effects of the intensification of agriculture on northern temperate dung beetle communities. *J. Appl. Ecol.* **2003**, *40*, 994–1007. [CrossRef]
20. Negro, M.; Rolando, A.; Palestrini, C. The Impact of Overgrazing on Dung Beetle Diversity in the Italian Maritime Alps. *Environ. Entomol.* **2011**, *40*, 1081–1092. [CrossRef]
21. Tonelli, M.; Verdu, J.R.; Zunino, M. Effects of the progressive abandonment of grazing on dung beetle biodiversity: Body size matters. *Biodivers. Conserv.* **2018**, *27*, 189–204. [CrossRef]

22. Tonelli, M.; Verdú, J.R.; Zunino, M. Grazing abandonment and dung beetle assemblage composition: Reproductive behaviour has something to say. *Ecol. Indic.* **2019**, *96*, 361–367. [CrossRef]

23. Seibold, S.; Gossner, M.M.; Simons, N.K.; Blüthgen, N.; Müller, J.; Ambarlı, D.; Ammer, C.; Bauhus, J.; Fischer, M.; Habel, J.C.; et al. Arthropod decline in grasslands and forests is associated with landscape-level drivers. *Nature* **2019**, *574*, 671–674. [CrossRef] [PubMed]

24. Geiger, F.; van der Lubbe, S.C.T.M.; Brunsting, A.M.H.; de Snoo, G.R. Insect abundance in cow dung pats of different farming systems. *Entomol. Ber.* **2010**, *70*, 106–110.

25. Lumaret, J.-P.; Errouissi, F. Use of anthelmintics in herbivores and evaluation of risks for the non target fauna of pastures. *Vet. Res.* **2002**, *33*, 547–562. [CrossRef] [PubMed]

26. Pecenka, J.R.; Lundgren, J.G. Effects of herd management and the use of ivermectin on dung arthropod communities in grasslands. *Basic Appl. Ecol.* **2019**, *40*, 19–29. [CrossRef]

27. Spector, S. Scarabaeine Dung Beetles (coleoptera: Scarabaeidae: Scarabaeinae): An Invertebrate Focal Taxon for Biodiversity Research and Conservation. *Coleopt. Bull.* **2006**, *60*, 71–83. [CrossRef]

28. Buse, J.; Entling, M.H. Stronger dung removal in forests compared with grassland is driven by trait composition and biomass of dung beetles. *Ecol. Entomol.* **2020**, *45*, 223–231. [CrossRef]

29. Numa, C.; Verdú, J.R.; Sánchez, A.; Galante, E. Effect of landscape structure on the spatial distribution of Mediterranean dung beetle diversity. *Divers. Distrib.* **2009**, *15*, 489–501. [CrossRef]

30. Roslin, T.; Viljanen, H. Dung Beetle Populations: Structure and Consequences. In *Ecology and Evolution of Dung Beetles*; Simmons, L.E., Ridsdill-Smith, T.J., Eds.; Wiley-Blackwell: Chichester, UK, 2011; pp. 220–244. ISBN 978-1-4443-4200-0.

31. Hartel, T.; Plieninger, T. *European Wood-Pastures in Transition: A Social-Ecological Approach*; Routledge: Abingdon, UK, 2014; ISBN 978-1-135-13911-7.

32. Plieninger, T.; Hartel, T.; Martín-López, B.; Beaufoy, G.; Bergmeier, E.; Kirby, K.; Montero, M.J.; Moreno, G.; Oteros-Rozas, E.; Van Uytvanck, J. Wood-pastures of Europe: Geographic coverage, social–ecological values, conservation management, and policy implications. *Biol. Conserv.* **2015**, *190*, 70–79. [CrossRef]

33. Stoate, C.; Báldi, A.; Beja, P.; Boatman, N.; Herzon, I.; Van Doorn, A.; De Snoo, G.; Rakosy, L.; Ramwell, C. Ecological impacts of early 21st century agricultural change in Europe—A review. *J. Environ. Manag.* **2009**, *91*, 22–46. [CrossRef]

34. Mihók, B.; Biró, M.; Molnár, Z.; Kovacs, E.; Bölöni, J.; Erős, T.; Standovár, T.; Török, P.; Csorba, G.; Margóczi, K.; et al. Biodiversity on the waves of history: Conservation in a changing social and institutional environment in Hungary, a post-soviet EU member state. *Biol. Conserv.* **2017**, *211*, 67–75. [CrossRef]

35. Varga, A.; Molnár, Z.; Biró, M.; Demeter, L.; Gellény, K.; Miókovics, E.; Molnár, Á.; Molnár, K.; Ujházy, N.; Ulicsni, V.; et al. Changing year-round habitat use of extensively grazing cattle, sheep and pigs in East-Central Europe between 1940 and 2014: Consequences for conservation and policy. *Agric. Ecosyst. Environ.* **2016**, *234*, 142–153. [CrossRef]

36. Hartel, T.; Hanspach, J.; Abson, D.; Máthé, O.; Moga, C.I.; Fischer, J. Bird communities in traditional wood-pastures with changing management in Eastern Europe. *Basic Appl. Ecol.* **2014**, *15*, 385–395. [CrossRef]

37. Gallé, R.; Urák, I.; Nikolett, G.-S.; Hartel, T. Sparse trees and shrubs confers a high biodiversity to pastures: Case study on spiders from Transylvania. *PLoS ONE* **2017**, *12*, e0183465. [CrossRef] [PubMed]

38. Da Silva, P.G.; Hernández, M.I.M. Spatial Patterns of Movement of Dung Beetle Species in a Tropical Forest Suggest a New Trap Spacing for Dung Beetle Biodiversity Studies. *PLoS ONE* **2015**, *10*, e0126112. [CrossRef] [PubMed]

39. Löbl, I.; Löbl, D. *Scarabaeoidea–Scirtoidea–Dascilloidea–Buprestoidea–Byrrhoidea: Revised and Updated Edition*; Catalogue of Palaearctic Coleoptera; Brill: Leiden, The Netherlands, 2016; ISBN 978-90-04-30914-2.

40. Zuur, A.; Ieno, E.N.; Walker, N.; Saveliev, A.A.; Smith, G.M. *Mixed Effects Models and Extensions in Ecology with R*; Statistics for Biology and Health; Springer: New York, NY, USA, 2009; ISBN 978-0-387-87457-9.

41. Chambers, J.M.; Hastie, T. *Statistical Models in S*; Wadsworth & Brooks/Cole Advanced Books & Software: Monterey, CA, USA, 1992; ISBN 978-0-534-16764-6.

42. Borcard, D.; Gillet, F.; Legendre, P. *Numerical Ecology with R*, 2nd ed.; Use R!; Springer International Publishing: New York, NY, USA, 2018; ISBN 978-3-319-71403-5.
43. R Core Team. R: A Language and Environment for Statistical Computing. 2018. Available online: https://www.r-project.org (accessed on 20 November 2020).
44. Bates, D.; Mächler, M.; Bolker, B.; Walker, S. Fitting Linear Mixed-Effects Models Using lme4. *J. Stat. Softw.* **2015**, *67*, 1–48. [CrossRef]
45. Venables, W.N.; Ripley, B.D. *Modern Applied Statistics with S*, 4th ed.; Statistics and Computing; Springer: New York, NY, USA, 2002; ISBN 978-0-387-95457-8.
46. Lenth, R.V. Least-Squares Means: The R Package lsmeans. *J. Stat. Softw.* **2016**, *69*, 1–33. [CrossRef]
47. Hothorn, T.; Bretz, F.; Westfall, P. Simultaneous Inference in General Parametric Models. *Biometr. J.* **2008**, *50*, 346–363. [CrossRef]
48. Oksanen, J.; Blanchet, F.G.; Friendly, M.; Kindt, R.; Legendre, P.; McGlinn, D.; Minchin, P.R.; O'Hara, R.B.; Simpson, G.L.; Solymos, P.; et al. Vegan: Community Ecology Package. 2019. Available online: https://CRAN.R-project.org/package=vegan (accessed on 20 November 2020).
49. Roberts, D.W. Labdsv: Ordination and Multivariate Analysis for Ecology. 2019. Available online: https://CRAN.R-project.org/package=labdsv (accessed on 20 November 2020).
50. Buse, J.; Šlachta, M.; Sladecek, F.X.J.; Carpaneto, G.M. Summary of the morphological and ecological traits of Central European dung beetles. *Entomol. Sci.* **2018**, *21*, 315–323. [CrossRef]
51. Ríos-Díaz, C.L.; Moreno, C.E.; Ortega-Martínez, I.J.; Zuria, I.; Escobar, F.; Castellanos, I. Sheep herding in small grasslands promotes dung beetle diversity in a mountain forest landscape. *J. Insect Conserv.* **2020**. [CrossRef]
52. Jugovic, J.; Koren, T.; Koprivnikar, N. Competition and Seasonal Co-Existence of Coprophagous Scarabaeoidea (Coleoptera) in Differently Managed Habitat Patches of Sub-Mediterranean Grasslands in Slovenia. *Pol. J. Ecol.* **2019**, *67*, 247–263. [CrossRef]
53. Numa, C.; Verdú, J.R.; Rueda, C.; Galante, E. Comparing Dung Beetle Species Assemblages Between Protected Areas and Adjacent Pasturelands in a Mediterranean Savanna Landscape. *Rangel. Ecol. Manag.* **2012**, *65*, 137–143. [CrossRef]
54. Tocco, C.; Negro, M.; Rolando, A.; Palestrini, C. Does natural reforestation represent a potential threat to dung beetle diversity in the Alps? *J. Insect Conserv.* **2013**, *17*, 207–217. [CrossRef]
55. Barton, P.S.; Manning, A.D.; Gibb, H.; Lindenmayer, D.B.; Cunningham, S.A. Conserving ground-dwelling beetles in an endangered woodland community: Multi-scale habitat effects on assemblage diversity. *Biol. Conserv.* **2009**, *142*, 1701–1709. [CrossRef]
56. Lövei, G.L.; Magura, T.; Tóthmérész, B.; Ködöböcz, V. The influence of matrix and edges on species richness patterns of ground beetles (Coleoptera: Carabidae) in habitat islands. *Glob. Ecol. Biogeogr.* **2006**, *15*, 283–289. [CrossRef]
57. Magura, T. Carabids and forest edge: Spatial pattern and edge effect. *For. Ecol. Manag.* **2002**, *157*, 23–37. [CrossRef]
58. Gallé, R.; Kanizsai, O.; Ács, V.; Molnár, B. Functioning of Ecotones—Spiders and Ants of Edges between Native and Non-Native Forest Plantations. *Pol. J. Ecol.* **2014**, *62*, 815–820. [CrossRef]
59. Bogyó, D.; Magura, T.; Nagy, D.D.; Tóthmérész, B. Distribution of millipedes (Myriapoda, Diplopoda) along a forest interior–forest edge–grassland habitat complex. *ZooKeys* **2015**, *510*, 181–195. [CrossRef]
60. Piccini, I.; Palestrini, C.; Rolando, A.; Roslin, T. Local management actions override farming systems in determining dung beetle species richness, abundance and biomass and associated ecosystem services. *Basic Appl. Ecol.* **2019**, *41*, 13–21. [CrossRef]
61. Bradford, M.A.; Berg, B.; Maynard, D.S.; Wieder, W.R.; Wood, S.A. Understanding the dominant controls on litter decomposition. *J. Ecol.* **2016**, *104*, 229–238. [CrossRef]
62. Von Hoermann, C.; Weithmann, S.; Deißler, M.; Ayasse, M.; Steiger, S. Forest habitat parameters influence abundance and diversity of cadaver-visiting dung beetles in Central Europe. *R. Soc. Open Sci.* **2020**, *7*, 191722. [CrossRef]

63. Perrin, W.; Moretti, M.; Vergnes, A.; Borcard, D.; Jay-Robert, P. Response of dung beetle assemblages to grazing intensity in two distinct bioclimatic contexts. *Agric. Ecosyst. Environ.* **2020**, *289*, 106740. [CrossRef]
64. Errouissi, F.; Jay-Robert, P. Consequences of habitat change in euromediterranean landscapes on the composition and diversity of dung beetle assemblages (Coleoptera, Scarabaeoidea). *J. Insect Conserv.* **2019**, *23*, 15–28. [CrossRef]
65. Dormont, L.; Rapior, S.; McKey, D.B.; Lumaret, J.-P. Influence of dung volatiles on the process of resource selection by coprophagous beetles. *Chemoecology* **2007**, *17*, 23–30. [CrossRef]
66. Wassmer, T. Seasonality of Coprophagous Beetles in the Kaiserstuhl Area near Freiburg (Sw Germany) Including the Winter Months. *Acta Oecol. Int. J. Ecol.* **1994**, *15*, 607–631.
67. Galante, E.; Mena, J.; Lumbreras, C. Dung Beetles (Coleoptera: Scarabaeidae, Geotrupidae) Attracted to Fresh Cattle Dung in Wooded and Open Pasture. *Environ. Entomol.* **1995**, *24*, 1063–1068. [CrossRef]
68. Agoglitta, R.; Moreno, C.E.; Zunino, M.; Bonsignori, G.; Dellacasa, M. Cumulative annual dung beetle diversity in Mediterranean seasonal environments. *Ecol. Res.* **2012**, *27*, 387–395. [CrossRef]
69. Senyuz, Y.; Lobo, J.M.; Dindar, K. Altitudinal gradient in species richness and composition of dung beetles (Coleoptera: Scarabaeidae) in an eastern Euro-Mediterranean locality: Functional, seasonal and habitat influences. *Eur. J. Entomol.* **2019**, *116*, 309–319. [CrossRef]
70. Piccini, I.; Caprio, E.; Palestrini, C.; Rolando, A. Ecosystem functioning in relation to species identity, density, and biomass in two tunneller dung beetles. *Ecol. Entomol.* **2020**, *45*, 311–320. [CrossRef]
71. Birkett, A.J.; Blackburn, G.A.; Menéndez, R. Linking species thermal tolerance to elevational range shifts in upland dung beetles. *Ecography* **2018**, *41*, 1510–1519. [CrossRef]
72. Storck-Tonon, D.; Da Silva, R.J.; Sawaris, L.; Vaz-De-Mello, F.Z.; Da Silva, D.J.; Peres, C.A. Habitat patch size and isolation drive the near-complete collapse of Amazonian dung beetle assemblages in a 30-year-old forest archipelago. *Biodivers. Conserv.* **2020**, *29*, 2419–2438. [CrossRef]
73. Jay-Robert, P.; Niogret, J.; Errouissi, F.; Labarussias, M.; Paoletti, É.; Luis, M.V.; Lumaret, J.-P. Relative efficiency of extensive grazing vs. wild ungulates management for dung beetle conservation in a heterogeneous landscape from Southern Europe (Scarabaeinae, Aphodiinae, Geotrupinae). *Biol. Conserv.* **2008**, *141*, 2879–2887. [CrossRef]
74. Buse, J.; Šlachta, M.; Sladecek, F.X.; Pung, M.; Wagner, T.; Entling, M.H. Relative importance of pasture size and grazing continuity for the long-term conservation of European dung beetles. *Biol. Conserv.* **2015**, *187*, 112–119. [CrossRef]
75. Forgie, S.A.; Paynter, Q.; Zhao, Z.; Flowers, C.; Fowler, S. Newly released non-native dung beetle species provide enhanced ecosystem services in New Zealand pastures. *Ecol. Entomol.* **2018**, *43*, 431–439. [CrossRef]
76. Vojta, J.; Drhovská, L. Are abandoned wooded pastures suitable refugia for forest species? *J. Veg. Sci.* **2012**, *23*, 880–891. [CrossRef]

Publisher's Note: MDPI stays neutral with regard to jurisdictional claims in published maps and institutional affiliations.

 *forests*

Article

Smaller and Isolated Grassland Fragments Are Exposed to Stronger Seed and Insect Predation in Habitat Edges

Kitti Kuli-Révész [1,2], Dávid Korányi [1], Tamás Lakatos [1,3], Ágota Réka Szabó [1,3], Péter Batáry [1] and Róbert Gallé [1,*]

[1] 'Lendület' Landscape and Conservation Ecology, Centre for Ecological Research, Institute of Ecology and Botany, H-2163 Vácrátót, Hungary; kuli-revesz.kitti@ecolres.hu (K.K.-R.); koranyi.david@ecolres.hu (D.K.); lakatos.tamas@ecolres.hu (T.L.); szabo.agota@ecolres.hu (Á.R.S.); batary.peter@ecolres.hu (P.B.)

[2] Doctoral School of Biological Sciences, Szent István University, H-2100 Gödöllő, Hungary

[3] Doctoral School of Biology, Eötvös Loránd University, H-1053 Budapest, Hungary

* Correspondence: galle.robert@ecolres.hu

Received: 23 November 2020; Accepted: 29 December 2020; Published: 2 January 2021

Abstract: Habitat fragmentation threatens terrestrial arthropod biodiversity, and thereby also leads to alterations of ecosystem functioning and stability. Predation on insects and seeds by arthropods are two very important ecological functions because of their community-structuring effects. We addressed the effect of fragment connectivity, fragment size, and edge effect on insect and seed predation of arthropods. We studied 60 natural fragments of two grassland ecosystems in the same region (Hungarian Great Plain), 30 forest-steppes, and 30 burial mounds (kurgans). The size of fragments were in the range of 0.16–6.88 ha for forest-steppe and 0.01–0.44 ha for kurgan. We used 2400 sentinel arthropod preys (dummy caterpillars) and 4800 seeds in trays for the measurements. Attack marks on dummy caterpillars were used for predator identification and calculation of insect predation rates. In the case of seeds, predation rates were calculated as the number of missing or damaged seeds per total number of exposed seeds. Increasing connectivity played a role only in generally small kurgans, with a negative effect on insect and seed predation rates in the edges. In contrast, fragment size moderated edge effects on insect and seed predation rates in generally large forest-steppes. The difference between edges and centres was more pronounced in small than in large fragments. Our study emphasizes the important role of landscape and fragment-scale factors interacting with edge effect in shaping ecosystem functions in natural grassland fragments of modified landscapes. Managing functional landscapes to optimize the assessment of ecosystem functions and services needs a multispatial scale approach.

Keywords: arthropod predation; connectivity; dummy caterpillar; ecosystem function; edge effect; forest-steppe; fragment size; kurgan; landscape-scale; seed predation

---

1. Introduction

Habitat loss and fragmentation are among the most relevant threats to arthropod biodiversity [1]. Agricultural expansion, afforestation with exotic tree species, and urbanization are the primary drivers of loss of natural or seminatural habitats and their insect communities [2], leading to small habitat fragments and decreased connectivity between them [3]. Classical island biogeography theory attempted to explain the effect of island size and distance from mainland sources on the diversity of species [4]. This concept was applied for terrestrial habitat fragments and the differences between oceanic islands, and isolated habitat fragments are now well-recognized [5,6]. The predictive power of habitat area was also demonstrated for arthropods of terrestrial islands [7,8].

The effect of decreasing connectivity on arthropods is highly taxon-specific. Habitat generalists and highly mobile species may cover large distances in a strongly modified landscape matrix [3]. The spatial proximity of suitable habitat fragments is more important for arthropods that are habitat specialists and have low mobility; thus, they may form isolated populations [9]. Furthermore, the conversion of a continuous habitat into disjunct habitat fragments usually increases the length of the edges between fragments and the surrounding matrix, which may significantly change the characteristics of edges, and the plant and animal diversity of communities [10,11].

Spillover is the movement of organisms across habitat edges [12]. Its effect is more pronounced near edges than in the central part of the habitat [13]. Most of the studies focused on how the influx of predators from seminatural habitats relates to the pest control services in agricultural fields [14–16]. Only a few studies found spillover from natural habitats [17,18]. For example, Madeira et al. [19] argue that spillover from adjacent crop habitats shapes carabid, rove beetle, and spider assemblages in fragmented seminatural grasslands.

Small habitat fragments are important biodiversity refuges [20] and may harbour a large proportion of the regional species pool in arable landscapes [21]. Species richness and density of arthropods in small fragments can be as high as in large ones [22,23]. However, there are some species that are disadvantaged in small habitats [24]. Changes in species richness and community composition can lead to alterations of ecosystem functioning and stability [25,26]; consequently, habitat fragmentation may broadly affect species interactions [27–29]. Furthermore, the effect of fragmentation on different ecosystem functions depends on the specific function and species identity [30]. Species of certain functional groups, such as larger body size or higher trophic level, may be more vulnerable to habitat loss, and this may have an effect on ecosystem functioning, resulting in a weaker top down effect in food webs [31]. However, the net effect of fragmentation remains controversial [32]. Large variation exists in how plant and animal species and species interactions respond to fragmentation. For example, Tong et al. [33] found that seed predation of acorn weevils (*Curculio glandium* Marsham) was high in large, less isolated fragments. In contrast, Elzinga et al. [34] found higher rates of seed predation on white campion (*Silene latifolia* Poir.) by the specialist moth lychnis (*Hadena bicruris* Hufnagel) in small fragments.

Insect and seed predation are important ecological functions because of the associated community-structuring effects [35,36]. Measuring species interactions such as insect and seed predation is challenging. Instead of measuring the function itself, studies often use densities of predators as a proxy [37], which can be misleading [38,39]. Here, we aimed to study the effects of fragmentation (i.e., increasing isolation, decreasing fragment size, and edge effect) directly on predation in two grassland ecosystems.

We chose forest-steppes and kurgans due to their intense exposure to fragmentation and their special role in nature conservation in the steppe zone [40]. Both types of steppe fragments have the potential to preserve the natural flora, fauna, and act as local biodiversity hotspots [40,41]. Forest-steppes are mosaics of grassland and forest fragments at the contact zone between closed-canopy temperate forests and steppe grasslands. They are among the most complex ecosystems in Eurasia, and their elements play a key role in landscape dynamics [41]. Kurgans (burial mounds) are artificial formations and were developed for burial purposes by steppic people (mainly in the range of IV–I millennia BC) by piling soil on the grave of an important person. The height of the kurgans ranges between half and a few meters, with the diameter between a few meters and 100 m [42]. These relatively small landscape elements represent important refuges for Eurasian steppe wildlife [43]. Both ecosystems are of high natural conservation value, harbouring numerous rare and protected plant and animal species. The fragment size and landscape structure of the two ecosystems are in different scales: small-scale landscape structure and relatively large fragment size in the case of forest-steppes, and large-scale landscape and small fragment size for kurgans.

However, the landscape matrix between fragments was relatively homogeneous and highly modified for both ecosystems. Our aim was to compare the two systems, and we expected different responses to the local and landscape factors.

We expected all studied fragmentation effects to be important determinants of insect and seed predation; however, the magnitude and relative importance of these effects, as well as their interaction, is not known. We tested the following hypotheses: (1) Predation rates are higher when connectivity decreases in the landscape, because isolation can enhance the spillover of generalist predators from the matrix. (2) Predation rates are higher in the edges than in the centres of a fragment, as a consequence of the edge effect. (3) Predation rates are lower in small than in large fragments, as functional groups of higher trophic levels are expected to be more sensitive to area loss. We aimed to reveal the similarities and differences of these questions in the two investigated fragmented grassland ecosystems of the same region using standardized methods.

2. Materials and Methods

2.1. Study Region and Sampling Design

We conducted our study on 60 natural grassland fragments in two different regions of the Hungarian Great Plain. We sampled 30 forest-steppe fragments in the central part of the Kiskunság region and 30 kurgans in southern Hungary. The investigated fragments were scattered around four settlements (Dévaványa, Kunágota, Makó, and Szentes) in the case of kurgans, and around three villages (Pirtó, Bócsa, and Kunfehértó) in the case of forest-steppes (Supplementary Material Figure S1). We established two transects of sentinel prey, and two trays of seeds in each centre and edge of every fragment (Figure 1B). Both areas are characterized by a continental climate with 500 to 550 mm mean annual precipitation, and 9.5 and 10 °C mean temperature, respectively [40,44]. Forest-steppes comprise extensive dry grasslands dominated by *Festuca vaginata* Waldst. and Kit ex Willd., *Stipa borysthenica* Klokov ex Prokudin, and relatively small forest fragments of poplar (*Populus alba* L.) and hawthorn (*Crataegus monogyna* Jacq.) [41]. Our study focused on dry steppic grasslands. The potential vegetation of kurgans consists of pannonic loess steppic grasslands [40] dominated by crested wheatgrass (*Agropyron cristatum* (L.) Gaertn.) and forage kochia *(Kochia prostrata* (L.) Schrad.) [45].

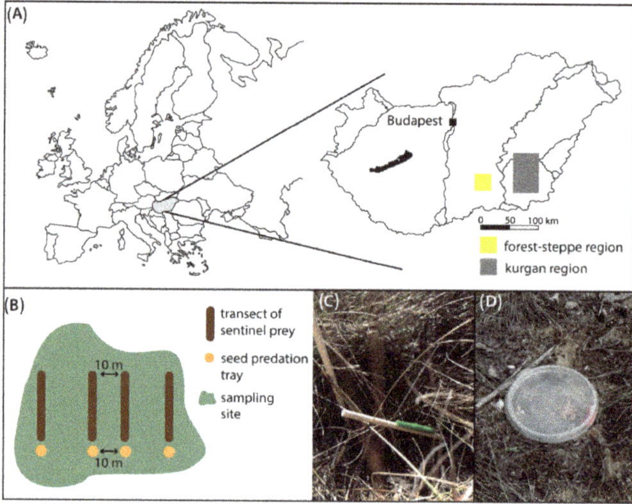

Figure 1. (**A**) Location of study regions in Hungary, Europe. (**B**) Sampling design. Light green represents the area of grassland fragment. Transects of sentinel preys and seed predation trays were minimum of 10 m away from each other, even in same transect position. (**C**) Sentinel prey. (**D**) Seed predation tray.

We selected the study sites on the basis of the size of the fragments and along a landscape configuration gradient by performing preliminary field visits and GIS calculations. We calculated Hanski's connectivity index [46] and hostile matrix percentage to quantify landscape configuration and composition using Google aerial photographs (captured in 2019), the basic ecosystem map of Hungary, and Quantum GIS 3.6.1 software [47]. Since kurgans and forest-steppes had two different spatial resolutions (i.e., kurgans were situated in large-scale agricultural landscapes and forest-steppe fragments were in a matrix of relatively small-scale forest plantations), we performed GIS calculations within a 1000 m radius buffer around the kurgans, and within a 500 m radius buffer around the forest-steppes. For connectivity calculations, we considered all habitat fragments (other forest-steppe fragments and open-sand grasslands for forest-steppes, closed and alkali grasslands for kurgans) that were located around the focal fragment. As we applied the connectivity index to entire predator communities containing many taxa, scaling parameters α and β were set to the value of 0.5 [48]. For hostile matrix calculations, we considered all nonhabitat fragments (coniferous and deciduous plantations, clear-cut areas, young afforestation for forest-steppes, and arable lands for kurgans) and calculated their pooled percentage cover in a buffer around each site. As we found significant correlations between hostile matrix percentage and connectivity in both habitat regions (forest-steppes: Pearson $r = -0.64$, $p < 0.001$; kurgans: Pearson $r = -0.95$, $p < 0.001$; i.e., proportion of hostile matrix significantly decreased with increasing connectivity), we used only connectivity as landscape-level variable in further analyses. Lastly, we selected 15 small (0.16–0.48 ha for forest-steppe; 0.01–0.10 ha for kurgan) and 15 large (0.93–6.88 ha for forest-steppe; 0.20–0.44 ha for kurgan) grassland fragments. Connectivity values of the selected fragments ranged from 0 (isolated) to 2637 (connected) for kurgans (mean = 689) and 24 to 811 for forest-steppes (mean = 394).

2.2. Sentinel Prey

We assessed the predatory activity of carnivorous insects with dummy green caterpillars of moths made of plasticine, exposed for seven days. This method of sentinel prey is easy to use and appropriate to assess in situ predation pressure [39,49]. Dummy caterpillars were 25 mm long and 5 mm in diameter, and made from light green nontoxic modelling plasticine (Fimo Soft®, Staedtler Mars GmbH & Co. KG, Nuremberg, Germany). All caterpillars were covered by PlastiDip® (PlastiDip International, Blaine, MN, USA) silicon spray to avoid drying and eliminate the smell of plasticine [50]. We fixed all caterpillars to 5 cm long wooden sticks with superglue for easier handling.

We attached them to the ground by pushing the end of the stick into the soil. We placed dummy caterpillars in transects, 1 m distance from each other. We used 2400 sentinel preys altogether (2 regions × 30 study sites × 2 transect positions × 2 transects × 10 caterpillars; Figure 1). The transects of sentinel preys were at a minimum of 10 m away from each other even in the same transect position. We installed dummy caterpillars on 21–27 June and collected them from 28 June to 4 July 2019. Potential predators were identified by the attack marks that they left on dummy caterpillars. We inspected the marks by using magnifying glasses and microscopes in the laboratory, following the methods described by Low et al. [51]. Multiple attack marks by the same predator group were assumed to originate from the same predator. Signs by different predator types were considered independent attacks.

2.3. Seed Predation

We exposed seeds in transparent, plastic trays to assess seed predation. Placing the seeds in shallow containers in the ground is a simple and established way for assessing seed predation [52,53]. We placed 10 seeds of *Triticum spelta* L. as large, and 10 seeds of *Festuca rubra* L. as small seeds in each tray. We used the different sizes to increase attractiveness for a wider range of seed predator arthropods. The trays were round plastic containers, 10 cm in diameter (Figure 1D). We fixed the container to the ground by attaching a plastic stick to the container and dug it into the soil. We excluded birds and rodents by closing the containers with transparent lids and creating 1 × 1 cm openings on their sides (only for arthropods). Altogether, we had 2 regions × 30 study sites × 2 transect positions × 2 trays, resulting in a total of 240 seed predation trays (Figure 1). The containers were a minimum of 10 m away from each other. We installed trays from 31 May to 6 June and collected them from 7 to 13 June 2019. Thus, all trays were exposed for 7 days. Seed predators were assumed to be responsible for missing seeds. We counted the remaining seeds in each tray and inspected them for further predation marks in the laboratory. We considered multiple attack marks on the same seed as one predation event. Several oligo- and monophagous specialist seed predator insects were present on our study sites, but their seed-predation effect was not included in our data.

2.4. Statistical Analysis

Insect predation rates were calculated as the number of sentinel prey items showing signs of predation per total number exposed per transect. Seed predation rates were calculated as the number of missing seeds and remaining seeds with predation marks per total seed number exposed per transect. To test whether connectivity, fragment size, transect position, and their second-order interactions (fixed factors) had a significant effect on insect and seed predation rates, we used generalized linear mixed-effects models with the model averaging method. Models were fitted with binomial distribution. Connectivity ranged between 0 and 1. We used lmer (lme4) [54] models with fragment ID within village as a nested random-effect term. We used seed size as an offset variable in models of seed predation rates. We calculated Akaike's information criteria corrected for small sample sizes (AICc) to rank candidate models. The models with <6

ΔAICc of the best model (i.e., the model with the lowest AICc) were used for model averaging [55,56] with the R package MuMIn [57].

3. Results

3.1. Sentinel Prey

Overall, 72.13% (1731/2400) of the dummy caterpillars were attacked. On the basis of the identification of attack marks, 87.52% of the predators were ants, 7.93% beetles, 5.21% reptiles, 3.40% wasps, 3.17% birds, 1.25% mammals, and 0.79% were bees. The effect of landscape and local variables was not unequivocal for the two ecosystems. Fragment size, transect position, and their interaction had the highest relative importance for insect predation in forest-steppes (Table 1). We detected higher predation rates in edges in the central transects of small forest-steppes; however, we detected the opposite pattern in large fragments (Figure 2A). We found no interaction effect between fragment size and transect position in kurgans (Table 1, Figure 2B). Variables that best explained insect predation in kurgans were connectivity, transect position, and their interaction according to relative importance values (Table 1). Increasing connectivity had a negative effect on the predation rates of edges but not in kurgan centres (Figure 2D).

Table 1. Summary table for GLMM results after multimodel averaging of the best candidate models showing relative importance of each explanatory variable (fragment size: large (L) vs. small (S), transect position: centre (C) vs. edge (E), and connectivity), and their interactions on insect and seed predation rates in forest-steppes (FS) and kurgans (KU).

Model [a]	Variable	Relative Importance (%) [b]	Multimodel Estimate ± 95% [c]	
Insect predation (FS)	Fragment size (L/S)	100	−0.215	±0.538
	Transect (C/E)	100	−0.449	±0.427 *
	Fragment size × transect	100	0.895	±0.538 **
	Connectivity	36	0.050	±0.928
	Connectivity × transect	8	0.300	±1.140
	Connectivity × fragment size	7	0.251	±1.689
Insect predation (KU)	Connectivity	67	−0.418	±1.887
	Transect (C/E)	62	0.405	±0.651
	Connectivity × Transect	49	−1.303	±1.014 *
	Fragment size (L/S)	33	−0.258	±1.087
	Fragment size × transect	4	−0.077	±0.573
	Connectivity × Fragment size	4	−0.425	±3.402
Seed predation (FS)	Fragment size (L/S)	100	−0.521	±0.651
	Transect (C/E)	100	0.208	±0.313
	Fragment size × Transect	100	0.635	±0.387 **
	Connectivity	38	0.071	±1.203
	Connectivity × transect	8	0.164	±0.827
	Connectivity × fragment size	8	−0.380	±2.125
Seed predation (KU)	Connectivity	100	1.281	±1.589
	Transect (C/E)	100	0.098	±0.439
	Connectivity × transect	100	−1.731	±0.872 ***
	Fragment size (L/S)	42	0.064	±1.037
	Fragment size × transect	15	0.227	±0.469
	Connectivity × Fragment size	11	−0.547	±2.797

[a] Models fitted with binomial distribution (number of candidate models, ΔAIC < 6). [b] Each variable's importance within best candidate models (ΔAIC < 6). [c] Significance levels: *: <0.05, **: <0.01, ***: <0.001.

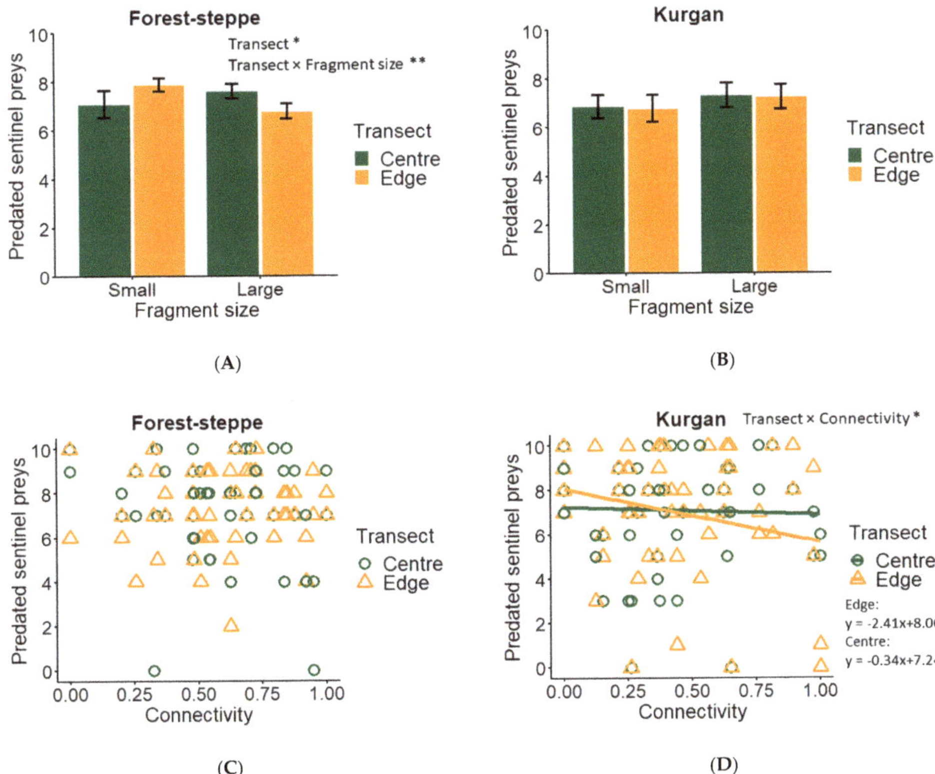

Figure 2. Insect predation. Interacting effect of (**A**) fragment size and edge effect on insect predation in forest-steppes; (**B**) fragment size and edge effect on insect predation in kurgans (mean ± SE); (**C**) connectivity and edge effect on insect predation in forest-steppes; and (**D**) connectivity and edge effect on insect predation in kurgans. Green, centre position; orange, edge position. Significance levels: *: <0.05, **: <0.01.

3.2. Seed Predation

In total, 77.58% (3724/4800) of the seeds had predation marks or been carried away. Similar to the results of insect predation, we found a significant interaction between fragment size and edge effect in case of forest-steppes, but not in kurgans (Table 1, Figure 3A,B). We found higher seed predation rates in edge in centre transects in forest-steppes, and this difference was more pronounced in small than in large fragments (Figure 3A). Connectivity and transect position did not affect seed predation in forest-steppes (Table 1, Figure 3C); however, seed predation in kurgans was affected by connectivity, transect position, and their interaction (Table 1, Figure 3D). In kurgans, we found a negative effect of connectivity on seed predation rates in edge and positive effect in centre transects (Figure 3D).

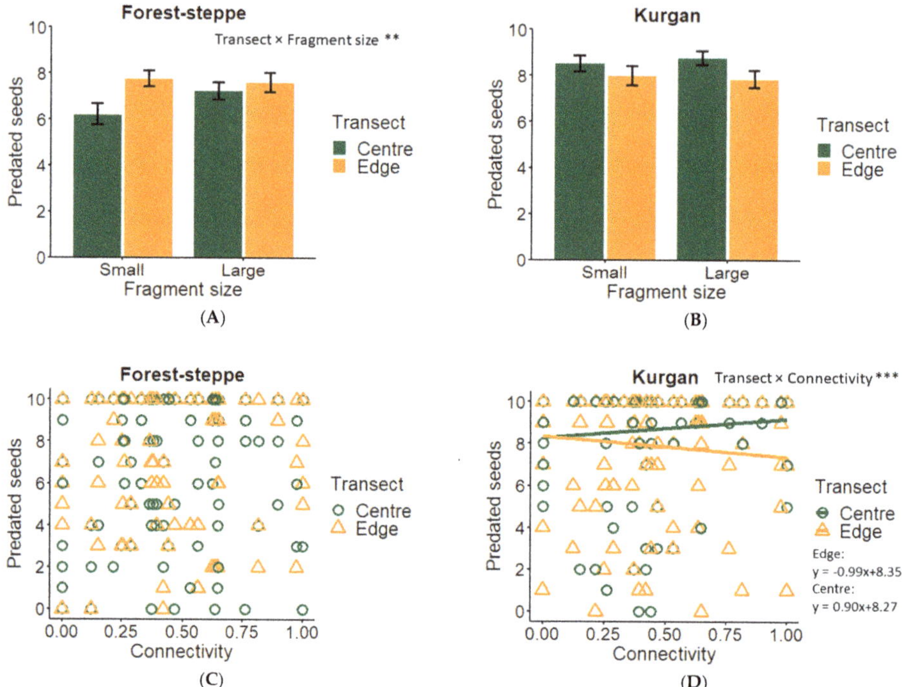

Figure 3. Seed predation. Interacting effect of (**A**) fragment size and edge effect on seed predation in forest-steppes; (**B**) fragment size and edge effect on seed predation in kurgans (mean ± SE); (**C**) connectivity and edge effect on seed predation in forest-steppes; and (**D**) connectivity and edge effect on seed predation in kurgans. Green, centre position; orange, edge position. Significance levels: **: <0.01, ***: <0.001.

4. Discussion

We addressed the effect of connectivity, fragment size, edge effect, and their interactions on insect and seed predation of arthropods. The predation rates of forest-steppes and kurgans responded differently to the effects of fragmentation depending on the fragmented grassland ecosystem. Predation pressure was higher in the edges of small fragments of forest-steppes and in the edges of more isolated landscapes of kurgans. Thereby, our results emphasize the differences of the two fragmented grassland ecosystems.

4.1. Predation in Forest-Steppe

Connectivity did not affect predation (Hypothesis 1); however, our second and third hypotheses about edge effect and fragment size, respectively, were partly supported, as the difference between edges and centres was more pronounced in small than in large fragments. In the case of insect predation, we found higher predation rates in the edges in the centres of small fragments, but we found the opposite pattern in the case of large fragments. Seed predation was higher in the edges of both small and large fragments, but the difference was more pronounced in small fragments.

Edge effects were stronger in small than in large fragments, which was in line with the findings of Laurance and Yensen [58]. This was explained partly by the higher edge/area ratio, i.e., smaller fragments had longer edges related to their area [59]. The edge effect on arthropod predation, and on other interspecific interactions and functional trait composition of arthropods, is relatively well-known [60–62]. Some of them measured with the very same sentinel prey method [40]. It is expected that resource consumption in terrestrial arthropod communities is higher at fragment edges [63]. The abundance of generalist predators increases near edges, which may alter food webs [64]. However, the response of insect predation by arthropods to edges is highly variable. Although there is a general positive correlation between the abundance of predators and predation rates [65,66], predation rates were found to be even higher in the habitat interior than in the edges in some cases [40]. There are contradictory results for seed removal ratios at edges vs. centres, with negative [67], or neutral [53] responses to edges. Fragmentation may shape many attributes of edges [10,11]; landscape and fragment-scale factors may interact with edge effects [63,67]. Our results emphasize that several effects may modify seed predation at edges, as we found higher rates at edges than in centres, but this effect was modified by fragment size. The interaction of these factors may explain the variability of predation responses to edges.

The main difference between forest-steppe and kurgan systems is the size of the fragments. Forest-steppe fragments (0.16–6.88 ha) were an order of magnitude larger kurgan fragments (0.01–0.44 ha), presumably maintained by abundant predator and seed predator assemblages. This might have had an effect on predation rates, which caused fragment size to override the effect of connectivity that we expected in the hypothesis (1).

4.2. Predation in Kurgan

Our first hypothesis that predation pressure increases with decreasing connectivity was supported for insect and seed predation in edges of kurgans. We found that increasing connectivity had a negative effect on predation rates in the edges but had a minor effect in the centres of kurgans. In addition, predation rates of edges mainly stayed under the rates of centres, in contrast to our presumptions during the hypothesis (2). Presumably, spillover from arable lands to grasslands increased predation rates in the edges.

We assumed that predators of arable fields dominated the predator assemblages in the edges of kurgans, therefore the decreasing amount of agricultural fields associated with increasing connectivity in the matrix had a negative effect on predation in the edges. Spillover between agricultural fields and natural habitats is recognised as an important mechanism shaping biodiversity, biotic interactions in communities, and ecosystem functions [19,68,69]. In order to gain insight into these mechanisms, not only should the number of species and individuals that move between adjacent habitats be observed, but also the functional responses should be addressed directly [70]. Native herbivores in natural habitats may experience increased predation pressure in landscapes with increasing habitat loss, due to spillover of generalist predators from surrounding cropland habitats [71]. Agricultural landscapes appear to augment generalist predators resulting in dramatically higher predator–prey ratios and reduced herbivore abundance [72].

Seed predation also decreased in the edges of kurgans when connectivity increased. Granivorous arthropods from the agricultural matrix may contribute greatly to seed predation rates in the edges of grasslands. When the proportion of arable fields decreased in the surrounding matrix; i.e., connectivity of habitat fragments increased, it led to a decrease in seed predation by arthropods. Our presumption about the tendency of movement from arable field to the adjacent habitat is recognised for some seed consumer carabids [73]. Furthermore, our results align with those of the study of Craig et al. [67], who found that seed predator arthropods cause higher predation rates with decreasing connectivity in the landscape. Taken together, connectivity had similar effects on insect and seed predation, presumably due to the same phenomenon: spillover.

Connectivity moderated the edge effect on predation; however, we did not find an effect of fragment size (Hypothesis 3), in contrast to in the forest-steppes. All kurgan fragments were relatively small.

Presumably, the effect of connectivity overrode the effect of fragment size. These results are in line with those of Rösch et al. [23], emphasising that connectivity is more important for small, isolated fragments, whereas fragment size is more relevant for large fragments.

5. Conclusions

Landscape and fragment-scale factors interact in shaping predation pressure by arthropods in natural grassland fragments. Thus, the edge effect was stronger in small forest-steppes and isolated kurgan fragments. Although our data were obtained only during a single year, our study assessing predation in seminatural grasslands improved our understanding of responses of predators to different local and landscape factors. Our study emphasizes that a multispatial scale approach is needed for the effective assessment of ecosystem functions, which may contribute to the conservation and maintenance of high-value seminatural grasslands.

Supplementary Materials: The following are available online at http://www.mdpi.com/1999-4907/12/1/54/s1, Figure S1: (a) Sampling sites on forest-steppes; (b) Sampling sites on kurgans. Aerial image was obtained from Google Earth.

Author Contributions: Conceptualization, P.B. and R.G.; methodology, R.G. and P.B.; formal analysis, D.K, P.B., R.G.; investigation, K.K.-R., Á.R.S., T.L., D.K.; resources, P.B.; writing—original-draft preparation, R.G. and K.K.-R.; writing—review and editing, P.B., Á.R.S., T.L., D.K; visualization, D.K., K.K.-R.; funding acquisition, P.B., R.G. All authors have read and agreed to the published version of the manuscript.

Funding: This research was funded by the Hungarian National Research, Development, and Innovation Office (NKFIH FK 131379 for R.G. and KKP 133839 for P.B.). R.G. was supported by the János Bolyai Research Scholarship of the Hungarian Academy of Sciences.

Acknowledgments: We are grateful to Nikolett Gallé-Szpisjak, Fabio Marcolin, Edina Törö, and Balázs Deák for their technical support. GIS analyses were performed using the Ecosystem BaseMap, Ministry of Agriculture, 2019 (KEHOP-430-VEKOP-15-2016-00001).

Conflicts of Interest: The authors declare no conflict of interest.

References

1. Cardoso, P.; Barton, P.S.; Birkhofer, K.; Chichorro, F.; Deacon, C.; Fartmann, T.; Fukushima, C.S.; Gaigher, R.; Habel, J.C.; Hallmann, C.A.; et al. Scientists' warning to humanity on insect extinctions. *Biol. Conserv.* **2020**, *242*, 108426. [CrossRef]
2. Habel, J.C.; Trusch, R.; Schmitt, T.; Ochse, M.; Ulrich, W. Long-term large-scale decline in relative abundances of butterfly and burnet moth species across south-western Germany. *Sci. Rep.* **2019**, *9*, 1–9. [CrossRef] [PubMed]
3. Fischer, J.; Lindenmayer, D.B. Landscape modification and habitat fragmentation: A synthesis. *Glob. Ecol. Biogeogr.* **2007**, *16*, 265–280. [CrossRef]
4. MacArthur, R.H.; Wilson, E.O. *The Theory of Island Biogeography*; Princeton University Press: Princeton, NJ, USA, 1967.
5. Laurance, W.F. Beyond island biogeography theory: Understanding habitat fragmentation in the real world. In *The Theory of Island Biogeography Revisited*; Losos, J.B., Ricklefs, R.E., Eds.; Princeton University Press: Princeton, NJ, USA, 2009; pp. 214–236.
6. Ord, T.J.; Emblen, J.; Hagman, M.; Shofner, R.; Unruh, S. Manipulation of habitat isolation and area implicates deterministic factors and limited neutrality in community assembly. *Ecol. Evol.* **2017**, *7*, 5845–5860. [CrossRef] [PubMed]
7. Baz, A.; Garcia-Boyero, A. The effects of forest fragmentation on butterfly communities in central Spain. *J. Biogeogr.* **1995**, 129–140. [CrossRef]

8. Franzén, M.; Schweiger, O.; Betzholtz, P.E. Species-area relationships are controlled by species traits. *PLoS ONE* **2012**, *7*, e37359. [CrossRef]
9. Hanski, I. *Metapopulation Ecology*; Oxford University Press Inc.: New York, NY, USA, 1999.
10. Murcia, C. Edge effects in fragmented forests: Implications for conservation. *Trends Ecol. Evol.* **1995**, *10*, 58–62. [CrossRef]
11. Ries, L.; Sisk, T.D. A predictive model of edge effects. *Ecology* **2004**, *85*, 2917–2926. [CrossRef]
12. Rand, T.A.; Tylianakis, J.M.; Tscharntke, T. Spillover edge effects: The dispersal of agriculturally subsidized insect natural enemies into adjacent natural habitats. *Ecol. Lett.* **2006**, *9*, 603–614. [CrossRef]
13. Boetzl, F.A.; Schneider, G.; Krauss, J. Asymmetric carabid beetle spillover between calcareous grasslands and coniferous forests. *J. Insect Conserv.* **2016**, *20*, 49–57. [CrossRef]
14. Bianchi, F.J.; Booij, C.J.H.; Tscharntke, T. Sustainable pest regulation in agricultural landscapes: A review on landscape composition, biodiversity and natural pest control. *Proc. R. Soc. B Biol. Sci.* **2006**, *273*, 1715–1727. [CrossRef] [PubMed]
15. Batáry, P.; Báldi, A.; Ekroos, J.; Gallé, R.; Grass, I.; Tscharntke, T. Biologia Futura: Landscape perspectives on farmland biodiversity conservation. *Biol. Futur.* **2020**, 1–10. [CrossRef]
16. Badenhausser, I.; Gross, N.; Mornet, V.; Roncoroni, M.; Saintilan, A.; Rusch, A. Increasing amount and quality of green infrastructures at different scales promotes biological control in agricultural landscapes. *Agric. Ecosyst. Environ.* **2020**, *290*, 106735. [CrossRef]
17. Blitzer, E.J.; Dormann, C.F.; Holzschuh, A.; Klein, A.M.; Rand, T.A.; Tscharntke, T. Spillover of functionally important organisms between managed and natural habitats. *Agric. Ecosyst. Environ.* **2012**, *146*, 34–43. [CrossRef]
18. Schneider, G.; Krauss, J.; Boetzl, F.A.; Fritze, M.A.; Steffan-Dewenter, I. Spillover from adjacent crop and forest habitats shapes carabid beetle assemblages in fragmented semi-natural grasslands. *Oecologia* **2016**, *182*, 1141–1150. [CrossRef]
19. Madeira, F.; Tscharntke, T.; Elek, Z.; Kormann, U.G.; Pons, X.; Rösch, V.; Samu, F.; Scherber, C.; Batáry, P. Spillover of arthropods from cropland to protected calcareous grassland–the neighbouring habitat matters. *Agric. Ecosyst. Environ.* **2016**, *235*, 127–133. [CrossRef]
20. Fahrig, L.; Arroyo-Rodríguez, V.; Bennett, J.R.; Boucher-Lalonde, V.; Cazetta, E.; Currie, D.J.; Eigenbrod, F.; Ford, A.T.; Harrison, S.P.; Jaeger, J.A.; et al. Is habitat fragmentation bad for biodiversity? *Biol. Conserv.* **2019**, *230*, 179–186. [CrossRef]
21. Riggi, L.G.; Berggren, Å. Small field islands systems include a large proportion of the regional orthopteran species pool in arable landscapes. *J. Insect Conserv.* **2020**, *24*, 695–703. [CrossRef]
22. Tscharntke, T.; Steffan-Dewenter, I.; Kruess, A.; Thies, C. Contribution of small habitat fragments to conservation of insect communities of grassland–cropland landscapes. *Ecol. Appl.* **2002**, *12*, 354–363. [CrossRef]
23. Rösch, V.; Tscharntke, T.; Scherber, C.; Batáry, P. Biodiversity conservation across taxa and landscapes requires many small as well as single large habitat fragments. *Oecologia* **2015**, *179*, 209–222. [CrossRef]
24. Zabel, J.; Tscharntke, T. Does fragmentation of Urtica habitats affect phytophagous and predatory insects differentially? *Oecologia* **1998**, *116*, 419–425. [CrossRef] [PubMed]
25. Tilman, D.; Isbell, F.; Cowles, J.M. Biodiversity and ecosystem functioning. *Annu. Rev. Ecol. Evol. Syst.* **2014**, *45*, 471–493. [CrossRef]
26. Rossetti, M.R.; Tscharntke, T.; Aguilar, R.; Batáry, P. Responses of insect herbivores and herbivory to habitat fragmentation: A hierarchical meta-analysis. *Ecol. Lett.* **2017**, *20*, 264–272. [CrossRef] [PubMed]
27. Brudvig, L.A.; Damschen, E.I.; Haddad, N.M.; Levey, D.J.; Tewksbury, J.J. The influence of habitat fragmentation on multiple plant–animal interactions and plant reproduction. *Ecology* **2015**, *96*, 2669–2678. [CrossRef]
28. Bagchi, R.; Brown, L.M.; Elphick, C.S.; Wagner, D.L.; Singer, M.S. Anthropogenic fragmentation of landscapes: Mechanisms for eroding the specificity of plant–herbivore interactions. *Oecologia* **2018**, *187*, 521–533. [CrossRef]
29. Liu, J.; Wilson, M.; Hu, G.; Liu, J.; Wu, J.; Yu, M. How does habitat fragmentation affect the biodiversity and ecosystem functioning relationship? *Landsc. Ecol.* **2018**, *33*, 341–352. [CrossRef]
30. Fleury, M.; Galetti, M. Forest fragment size and microhabitat effects on palm seed predation. *Biol. Conserv.* **2006**, *131*, 1–13. [CrossRef]

31. González-Fernández, J.; De la Peña, F.; Hormaza, J.; Boyero, J.; Vela, J.; Wong, E.; Trigo, M.M.; Montserrat, M. Alternative food improves the combined effect of an omnivore and a predator on biological pest control. A case study in avocado orchards. *Bull. Entomol. Res.* **2009**, *99*, 433–444. [CrossRef]

32. Garcia, D.; Chacoff, N.P. Scale-dependent effects of habitat fragmentation on hawthorn pollination, frugivory, and seed predation. *Conserv. Biol.* **2007**, *21*, 400–411. [CrossRef]

33. Tong, X.; Zhang, Y.X.; Wang, R.; Inbar, M.; Chen, X.Y. Habitat fragmentation alters predator satiation of acorns. *J. Plant Ecol.* **2017**, *10*, 67–73. [CrossRef]

34. Elzinga, J.A.; Turin, H.; Van Damme, J.M.; Biere, A. Plant population size and isolation affect herbivory of Silene latifolia by the specialist herbivore Hadena bicruris and parasitism of the herbivore by parasitoids. *Oecologia* **2005**, *144*, 416–426. [CrossRef] [PubMed]

35. Eötvös, C.B.; Magura, T.; Lövei, G.L. A meta-analysis indicates reduced predation pressure with increasing urbanization. *Landsc. Urban Plan.* **2018**, *180*, 54–59. [CrossRef]

36. Auld, T.D.; Denham, A.J. The role of ants and mammals in dispersal and post-dispersal seed predation of the shrubs Grevillea (Proteaceae). *Plant Ecol.* **1999**, *144*, 201–213. [CrossRef]

37. Rusch, A.; Birkhofer, K.; Bommarco, R.; Smith, H.G.; Ekbom, B. Predator body sizes and habitat preferences predict predation rates in an agroecosystem. *Basic Appl. Ecol.* **2015**, *16*, 250–259. [CrossRef]

38. Lövei, G.L.; Ferrante, M. A review of the sentinel prey method as a way of quantifying invertebrate predation under field conditions. *Insect Sci.* **2017**, *24*, 528–542. [CrossRef]

39. Imboma, T.S.; Gao, D.P.; You, M.S.; You, S.; Lövei, G.L. Predation Pressure in Tea (*Camellia sinensis*) Plantations in Southeastern China Measured by the Sentinel Prey Method. *Insects* **2020**, *11*, 212. [CrossRef]

40. Deák, B.; Valkó, O.; Török, P.; Tóthmérész, B. Factors threatening grassland specialist plants—A multi-proxy study on the vegetation of isolated grasslands. *Biol. Conserv.* **2016**, *204*, 255–262. [CrossRef]

41. Erdős, L.; Tölgyesi, C.; Horzse, M.; Tolnay, D.; Hurton, Á.; Schulcz, N.; Körmöczi, L.; Lengyel, A.; Bátori, Z. Habitat complexity of the Pannonian forest-steppe zone and its nature conservation implications. *Ecol. Complex.* **2014**, *17*, 107–118. [CrossRef]

42. Deák, B.; Tóthmérész, B.; Valkó, O.; Sudnik-Wójcikowska, B.; Bragina, T.M.; Moysiyenko, I.I.; Bragina, T.M.; Apostolova, I.; Dembicz, I.; Bykov, N.I. Cultural monuments and nature conservation: The role of kurgans in maintaining steppe vegetation. *Biodivers. Conserv.* **2016**, *25*, 2473–2490. [CrossRef]

43. Deák, B.; Tóth, C.A.; Bede, Á.; Apostolova, I.; Bragina, T.M.; Báthori, F.; Bán, M. Eurasian Kurgan Database–a citizen science tool for conserving grasslands on historical sites. *Hacquetia* **2019**, *18*, 179–187. [CrossRef]

44. Fekete, G.; Molnár, Z.; Kun, A.; Botta-Dukát, Z. On the structure of the Pannonian forest-steppe: Grasslands on sand. *Acta Zool. Acad. Sci. Hung.* **2002**, *48* (Suppl. 1), 137–150.

45. Bölöni, J.; Molnár, Z.; Kun, A.; Biró, M. *Általános Nemzeti Élőhely-Osztályozási Rendszer (Á-NÉR 2007)*; MTA ÖBKI: Vácrátót, Hungary, 2007; p. 184.

46. Hanski, I.; Ovaskainen, O. The metapopulation capacity of a fragmented landscape. *Nature* **2000**, *404*, 755–758. [CrossRef] [PubMed]

47. Quantum GIS Development Team. Quantum GIS Geographic Information System. Open Source Geospatial Foundation Project. 2019. Available online: http://qgis.osgeo.org (accessed on 10 March 2019).

48. Kormann, U.; Rösch, V.; Batáry, P.; Tscharntke, T.; Orci, K.M.; Samu, F.; Scherber, C. Local and landscape management drive trait-mediated biodiversity of nine taxa on small grassland fragments. *Divers. Distrib.* **2015**, *21*, 1204–1217. [CrossRef]

49. Howe, A.; Lövei, G.L.; Nachman, G. Dummy caterpillars as a simple method to assess predation rates on invertebrates in a tropical agroecosystem. *Entomol. Exp. Appl.* **2009**, *131*, 325–329. [CrossRef]

50. Purger, J.J.; Kurucz, K.; Tóth, Á.; Batáry, P. Coating plasticine eggs can eliminate the overestimation of predation on artificial ground nests. *Bird Study* **2012**, *59*, 350–352. [CrossRef]

51. Low, P.A.; Sam, K.; McArthur, C.; Posa, M.R.C.; Hochuli, D.F. Determining predator identity from attack marks left in model caterpillars: Guidelines for best practice. *Entomol. Exp. Appl.* **2014**, *152*, 120–126. [CrossRef]

52. Brown, J.H.; Grover, J.J.; Davidson, D.W.; Lieberman, G.A. A preliminary study of seed predation in desert and montane habitats. *Ecology* **1975**, *56*, 987–992. [CrossRef]

53. Linabury, M.C.; Turley, N.E.; Brudvig, L.A. Insects remove more seeds mammals in first-year prairie restorations. *Restor. Ecol.* **2019**, *27*, 1300–1306. [CrossRef]
54. Bates, D.; Maechler, M.; Bolker, B. Walker, S. Fitting linear mixed-effects models using lme4. *J. Stat. Softw.* **2015**, *67*, 1–48. [CrossRef]
55. Richards, S.A. Dealing with overdispersed count data in applied ecology. *J. Appl. Ecol.* **2008**, *45*, 218–227. [CrossRef]
56. Bolker, B.M.; Brooks, M.E.; Clark, C.J.; Geange, S.W.; Poulsen, J.R.; Stevens, M.H.H.; White, J.S.S. Generalized linear mixed models: A practical guide for ecology and evolution. *Trends Ecol. Evol.* **2009**, *24*, 127–135. [CrossRef] [PubMed]
57. Barton, K. MuMIn: R Functions for Model Selection and Model Averaging. R Package Version 0.12.0. 2009. Available online: http://r-forge.r-project.org/projects/mumin (accessed on 2 June 2020).
58. Laurance, W.F.; Yensen, E. Predicting the impacts of edge effects in fragmented habitats. *Biol. Conserv.* **1991**, *55*, 77–92. [CrossRef]
59. Orrock, J.L.; Damschen, E.I. Corridors cause differential seed predation. *Ecol. Appl.* **2005**, *15*, 793–798. [CrossRef]
60. Murphy, S.M.; Battocletti, A.H.; Tinghitella, R.M.; Wimp, G.M.; Ries, L. Complex community and evolutionary responses to habitat fragmentation and habitat edges: What can we learn from insect science? *Curr. Opin. Insect Sci.* **2016**, *14*, 61–65. [CrossRef]
61. Wimp, G.M.; Ries, L.; Lewis, D.; Murphy, S.M. Habitat edge responses of generalist predators are predicted by prey and structural resources. *Ecology* **2019**, *100*, e02662. [CrossRef]
62. Gallé, R.; Geppert, C.; Földesi, R.; Tscharntke, T.; Batáry, P. Arthropod functional traits shaped by landscape-scale field size, local agri-environment schemes and edge effects. *Basic Appl. Ecol.* **2020**, *48*, 102–111. [CrossRef]
63. Martinson, H.M.; Fagan, W.F. Trophic disruption: A meta-analysis of how habitat fragmentation affects resource consumption in terrestrial arthropod systems. *Ecol. Lett.* **2014**, *17*, 1178–1189. [CrossRef]
64. Wimp, G.M.; Murphy, S.M.; Lewis, D.; Ries, L. Do edge responses cascade up or down a multi-trophic food web? *Ecol. Lett.* **2011**, *14*, 863–870. [CrossRef]
65. Ries, L.; Fagan, W.F. Habitat edges as a potential ecological trap for an insect predator. *Ecol. Entomol.* **2003**, *28*, 567–572. [CrossRef]
66. Thomson, L.J.; Hoffmann, A.A. Spatial scale of benefits from adjacent woody vegetation on natural enemies within vineyards. *Biol. Control* **2013**, *64*, 57–65. [CrossRef]
67. Craig, M.T.; Orrock, J.L.; Brudvig, L.A. Edge-mediated patterns of seed removal in experimentally connected and fragmented landscapes. *Landsc. Ecol.* **2011**, *26*, 1373–1381. [CrossRef]
68. McCoy, M.W.; Barfield, M.; Holt, R.D. Predator shadows: Complex life histories as generators of spatially patterned indirect interactions across ecosystems. *Oikos* **2009**, *118*, 87–100. [CrossRef]
69. Tölgyesi, C.; Császár, P.; Torma, A.; Török, P.; Bátori, Z.; Gallé, R. Think twice before using narrow buffers: Attenuating mowing-induced arthropod spillover at forest–grassland edges. *Agric. Ecosyst. Environ.* **2018**, *255*, 37–44. [CrossRef]
70. Schneider, G.; Krauss, J.; Steffan-Dewenter, I. Predation rates on semi-natural grasslands depend on adjacent habitat type. *Basic Appl. Ecol.* **2013**, *14*, 614–621. [CrossRef]
71. Rand, T.A.; Louda, S.M. Spillover of agriculturally subsidized predators as a potential threat to native insect herbivores in fragmented landscapes. *Conserv. Biol.* **2006**, *20*, 1720–1729. [CrossRef] [PubMed]
72. Rand, T.A.; Tscharntke, T. Contrasting effects of natural habitat loss on generalist and specialist aphid natural enemies. *Oikos* **2007**, *116*, 1353–1362. [CrossRef]

73.   Labruyere, S.; Petit, S.; Ricci, B. Annual variation of oilseed rape habitat quality and role of grassy field margins for seed eating carabids in arable mosaics. *Agric. For. Entomol.* **2018**, *20*, 234–245. [CrossRef]

Publisher's Note: MDPI stays neutral with regard to jurisdictional claims in published maps and institutional affiliations.

MDPI

St. Alban-Anlage 66

4052 Basel

Switzerland

Tel. +41 61 683 77 34

Fax +41 61 302 89 18

www.mdpi.com

Forests Editorial Office

E-mail: forests@mdpi.com

www.mdpi.com/journal/forests